The Technology to
Accounting Software

A Handbook for
Evaluating Vendor Applications

Stewart McKie

A Division of
DUKE COMMUNICATIONS INTERNATIONAL

221 E. 29th Street • Loveland, CO 80538
(800) 621-1544 • (970) 663-4700 • www.dukepress.com

Library of Congress Cataloging-in-Publication Data

McKie, Stewart.
    The technology guide to accounting software / by Stewart McKie.
        p. cm.
    Includes index.
    ISBN 1-882419-55-3 (pbk.)
    1. Accounting—Computer networks. 2. Client/server computing.
    3. Accounting—Computer programs. 4. Communications software.
    I. Title
    HF5625.7.M33 1997
    657'.0285'53—dc21   97-4605
                                                                        CIP

Copyright © 1997 by DUKE PRESS
DUKE COMMUNICATIONS INTERNATIONAL
Loveland, Colorado

This book was printed and bound in the United States of America.

ISBN 1-882419-55-3

1 2 3 4 5 6 EB 9 8 7

I dedicate this book to my parents and to my wife, Theresa.

Acknowledgments

This book in part incorporates and builds upon a series of articles I wrote for *Controller Magazine* in my role as contributing technology editor in 1995 and 1996. I would like to thank Greg Northrup, publisher of *Controller*, for the opportunity to cover these topics in the magazine and Laurie Brannen and the rest of the *Controller* editorial staff for their help in shaping the articles for publication.

I would also like to thank the many other people who have influenced my views and added value along the way, including subscribers to my newsletter, CFO/Info, my white-paper clients, and the many people in the accounting software industry I talk to on a daily basis. I hope the content of the book reflects some of their opinions and views. Thanks also to my editor at Duke Press, Barb Gibbens, for her attention to detail and the many useful suggestions she made to make the text more readable.

I wrote the manuscript for this book using Microsoft Word 6.0 (Microsoft Corporation, Redmond, Washington) and used symbols from Visio 4.0 (Visio Corporation, Seattle, Washington) in the diagrams. The screenshots I've used to illustrate concepts were provided by the vendors of the product shown.

My use of screenshots does not imply endorsement of vendors or products. I use trademarks and tradenames extensively and recognize the right of the trademark and tradename owners to these marks and names.

Table of Contents
At a Glance

Table of Contents

Chapter 10: Component Accounting

Part 2: Evaluating Accounting Software

Preface

The Technology Guide to Accounting Software: A Handbook for Evaluating Vendor Applications is written for managers involved in evaluating financial accounting software. The book aims to help these managers understand and evaluate accounting software from a technology perspective. The book makes accounting software evaluation easier by helping managers cut through the marketing hype and by providing a range of checklists for quickly short-listing products to evaluate in more detail.

You will find this book helpful for your next accounting software evaluation project if you are

- a CFO, controller, or financial system manager

- a business analyst or information systems analyst

- a practicing accountant advising clients

- a consultant or software reseller

You can evaluate accounting software from at least three perspectives.

- you can evaluate the software vendor, including the vendor's track record, financial stability, customer base, distribution, support policies, and vision.

- You can evaluate the application's functionality and how closely it fits the needs of your current and future business processes.

- You can evaluate the software technology and whether the product's technological foundation is sound, properly utilized, and able to deliver real business benefit.

This book focuses on the technology perspective. In modern accounting software, technology is key to the delivery of functional value. Yet many accounting system managers are not as familiar with the technology as they need to be to make informed buying decisions. Technology changes so quickly that it's all too easy for managers to ignore how effectively an application uses technology and choose instead to limit their investigation to the comfortable realm of functionality.

Many managers review accounting software only from a functional perspective — what the software does, rather than how it does it — and leave evaluation of the technology to others. I believe this approach is no longer acceptable. The technology a package uses determines not only the functionality it provides, but also whether that functionality will be able to sustain growth, accommodate change, and allow the enterprise to maintain its edge by incorporating new technologies as they become available. Before you can evaluate an accounting application from a functional perspective, you must have a clear understanding of its technological foundation.

To make understanding and evaluating accounting software technology easier, I've divided this book into two parts. Part 1 discusses 10 technology topics that affect the design, deployment, and functionality of accounting software. A basic overview of the topic describes the technology and its capabilities. The rest of the chapter discusses how the technology is being used or is likely to be used in the future and the effect the technology is likely to have on business and answers specific questions that are often asked regarding that technology. Some chapters include screenshots that illustrate how products deliver specific functionality.

Near the beginning of each chapter in Part 1 is an "Evaluation Heads-Up" section that summarizes the main points of the chapter from an evaluation perspective. A "Questions for Vendors" box at the end will help you ask the right questions during your evaluation.

Part 2 is designed to help you produce a short-list of vendors and products that meet your needs. Chapter 11 segments the accounting-software market into four classes of solutions and lists key differentiators to help you distinguish between and rank what can sometimes be dozens of similar systems. Chapter 12 suggests 20 questions to ask vendors as part of your initial shortlisting process and provides a simple weighting system for the responses. Chapter 13 provides 50 leading accounting vendors in an easy-to-read, one-page overview of each vendor and its products. The profiles reflect many of the technologies discussed in Part 1. Finally, Chapter 14 provides nine cross-reference lists that help you compare products by their technologies and features.

Chapter 1

Client/Server Architectures

Client/Server 101

Client/server architecture differs from the terminal/host and PC/file server architectures that preceded it, but its roots are in those older architectures. Applications designed for client/server deployment are also ideally suited for adapting to the new world of Internet and intranet computing. To understand the client/server architecture's place in modern business and the advantages it provides, you must understand the architectures that preceded it.

Terminal/Host Architecture

Figure 1.1 shows a timeline of business computing architectures. The early years were dominated by the terminal/host computing architecture, in which the host — originally a mainframe; later a minicomputer or Unix server — centralizes all aspects of the business processing of an accounting application. Users interact with the accounting application through a "dumb" terminal — a visual display unit connected to the host either locally or remotely. The host manages the terminal display, which acts simply as a window to application software that runs on the host processor.

Terminals have no usable local resources (chips, memory, or disk) for processing data; rather, they act as visual "punch cards" for the user to enter, request, and display data. Even when the terminal is actually a PC that's simply emulating a terminal, there's no difference in the way the terminal/host architecture works. Despite the fact that the terminal/host architecture is still the most widely used transaction processing platform in the world, as a platform for end-user-oriented business information processing, it has a number of drawbacks. Some of these drawbacks are

Figure 1.1: Timeline of Business Computing Architectures

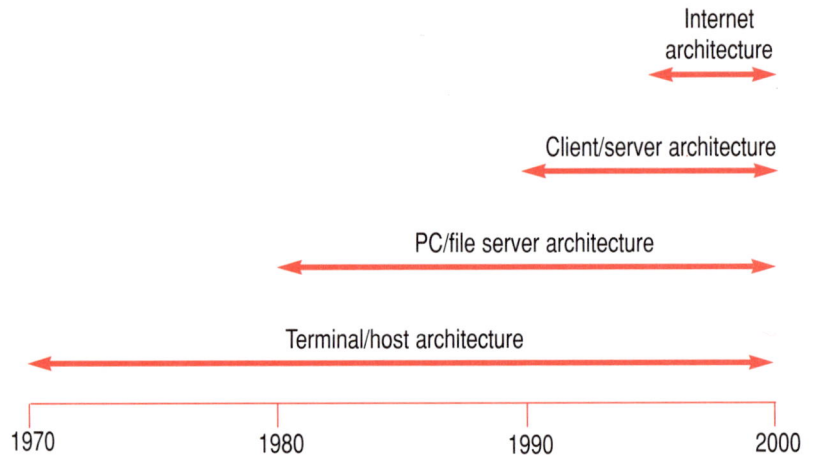

- Data the user enters must be dispatched to the host and validated, as few host accounting applications offer context-sensitive error checking and help online.

- Dumb terminals can support only nongraphical applications, which are less attractive to users and lack many ease-of-use features.

- When all processing is centralized on the host, downtime and spikes in user loads or activity slow the system down for all connected users.

- The complexity of host-based software means that mainframe software vendors are slow to enhance their software and incorporate new technology into their applications.

- Application users are locked in to one platform because accounting software is designed to work on specific host processors and is seldom portable to other processors or platforms.

- Host-based computers, peripherals, and software are expensive to acquire and maintain compared to other platforms and often require special environmental operating conditions.

- Host-system scalability is limited to the family of computers to which the host belongs, thus limiting the granularity of scalability options and increasing upgrade costs.

EVALUATION HEADS-UP

- Accounting applications designed for host- or file-server-based architectures don't migrate well to client/server, nor do they take full advantage of the architecture. If you intend to leverage your accounting system over a five- to ten-year life cycle, you should consider only applications designed specifically for client/server.

- Properly partitioned client/server accounting applications allow a choice of mixed client and server deployments that make implementing a system across headquarters, regional, or business-unit sites easier to manage.

- Because the server is so important for application and transaction management in client/server architectures, you should allocate most of your time and effort to selecting the right server and server operating system for the accounting system's deployment needs.

- An overload of untried and untested system components can hold up the transition to a complex client/server application. Make sure you have a fast and stable network infrastructure in place before you implement your chosen package.

- Review your current and anticipated future processing needs from a transaction, user, and functional perspective to determine whether you can live with two-tier client/server accounting applications or whether you should instead be looking for three- or n-tier accounting applications, which offer better all-round scalability.

- Do not expect client/server accounting applications to necessarily match the functional depth and breadth of your current applications, which may have benefited from a 10- to 20-year development life cycle. On the other hand, you can expect these new packages to be enhanced and upgraded faster than were previous generations of accounting software.

PC/File Server Architecture

When the PC was introduced for business use in the early 1980s, it brought with it a fundamental paradigm shift: for the first time, computing resources available on users' desktops let them process data locally instead of only on a remote host. With the introduction of local area networking software in the mid 1980s, local processing could be extended to include sharing files stored on networked PC file servers accessible by workgroups of users.

Many small-business and corporate departmental accounting systems could now be implemented on Novell NetWare-, IBM PC Network-, or Microsoft LAN Manager-based local area networks. The accounting data was stored in files on file servers and managed using simple file management software such as Cobol ISAM (indexed sequential access method) or Btrieve. The biggest problem with the file server architecture was that a copy of the file to be shared had to be transferred to the user's PC so it could process the data in the file locally. As a result, from an accounting application perspective, the file server approach to managing accounting data and applications also has some drawbacks:

- Network traffic can increase and system performance can degrade as data files grow and user demand increases.

- File management software is relatively unsophisticated in terms of security, reliability, and administrative functionality.

- Some file systems operate their locking at the file level, causing users to be locked out of a file whenever it's checked out by another user.

Client/Server Architecture

The client/server architecture has the potential to eliminate many of the drawbacks of both the terminal/host and PC/file server architectures. Client/server is based on a collection of client and server computers and other devices, such as telephones and fax machines, all connected by local area network (LAN) or wide area network (WAN) infrastructures. Client/server is fundamentally a network computing architecture that is not based on the use of any one processing architecture or any one vendor-specific computer platform. Conceptually, client devices request data from and submit it to servers, which in turn manage, process, and return data to clients across the network.

However, client/server roles are fluid: clients can act as servers to other clients, and servers can act as clients to other servers. Clients may be many types of devices, and servers may be based on many types of computer, as we'll see. From an application perspective, the most important aspect of client/server is that it encourages application developers to partition their applications into components that can be deployed across clients and servers, rather than running whole applications either on the server (as in host-based processing) or on the client (as in file server-based processing). Consequently, the client/server architecture offers a number of potential advantages over previous computing architectures. Some of these are

- Client/server is platform-neutral, allowing client and server platforms to be mixed and matched to suit the business need and budget.

- Client/server is highly scalable because the client, server, and network components can all be sized individually in response to changes in system demand.

- Client/server is highly adaptable — you can "plug-in" new client devices and servers to adapt to new technology, such as telephony or the Internet.

- Client/server technology can embrace host and file-server applications.

- Because client/server can absorb new technologies, such as the Internet and component-based software, it offers architectural longevity.

- Client/server can take full advantage of decreasing memory and disk costs and of increasing network bandwidth technology for improving overall and component-system performance.

Of course, client/server also has drawbacks:

- Client/server development tools and expertise are relatively immature compared to mainframe equivalents that have been refined over two decades.

- Client/server architectures are more complex to conceptualize, implement, and manage effectively, making life more challenging for IS departments.

- Because many client/server accounting applications are new to market or only in their first generation of release, they may not be fully robust or offer the functional breadth and depth of previous generations of applications.

- The intangible costs of client/server (e.g., retraining, implementation consulting, fixing problems, poor support) mean that it's unlikely to be a cheap alternative to existing architectures.

- Because the main benefit of client/server computing is reckoned to be improvements in productivity, return-on-investment data is hard to collect, and the return is difficult to prove.

However, many of the drawbacks are due to the immaturity of client/server tools, expertise, and implementations, not to any fundamental flaw in the architecture's conceptual basis. And the drawbacks have not stopped thousands of organizations worldwide from embracing client/server accounting, either as "downsizers" or "rightsizers" from mainframe or midrange systems or "upsizers" from PC LAN-based accounting systems.

What's Different About Client/Server Applications?

Accounting applications designed for use on client/server architectures must be written specifically for the architecture to take advantage of it. Although you can port host and file-server applications to and run them on client/server architectures, there is little point in doing so because the applications can't take advantage of the benefits client/server offers. This need for a complete redesign is why almost every leading accounting software vendor worldwide has produced new or rewritten applications specifically for use on the client/server architecture. (For a list of many of these vendors and their products, see Chapter 13.)

In practice, a commercial client/server application can be characterized as having

- a graphical client or "front-end" application that runs on an Intel- or RISC-based desktop workstation for data entry, retrieval, and analysis

- a server-based or "back-end" relational database management system (RDBMS) that runs on powerful single- or multiprocessor computers for managing the data

- a LAN or WAN infrastructure for communicating and sharing data across departmental or enterprise workgroups

But in fact, none of these attributes are necessary for an application to be termed "client/server." To understand the real difference about client/server applications, it helps to understand the three layers that make up the basic architecture of most business applications (Figure 1.2).

- Presentation services are the functions that present the screens (or forms) used to enter, query, display, or manipulate data.

- Application services provide the rules and logic for managing data entry, validating data, handling errors, providing user help and information messages, and doing complex business-specific processing such as posting transactions.

- Database services provide the data storage and database management tools for maintaining the data.

Rather than being monolithic, a well-designed client/server application is partitioned so that some or all of these three functional layers can be deployed separately or in combination on both clients and servers. An application partitioned in this way can be deployed in a wide variety of configurations compared to applications designed for use on host- or file server-based architectures. Partitioning lets businesses implement client/server accounting differently at, say, headquarters, where more computing resources may be available, than they do at the level of individual business units. This software design granularity also provides the foundation for client/server's scalability, because layers can be "scaled up" (run on a more powerful computer) independently of one another. So if transaction volume increases, for example, you can increase the power of your database server while leaving the rest of the system unchanged.

Figure 1.2: Business Application Layers

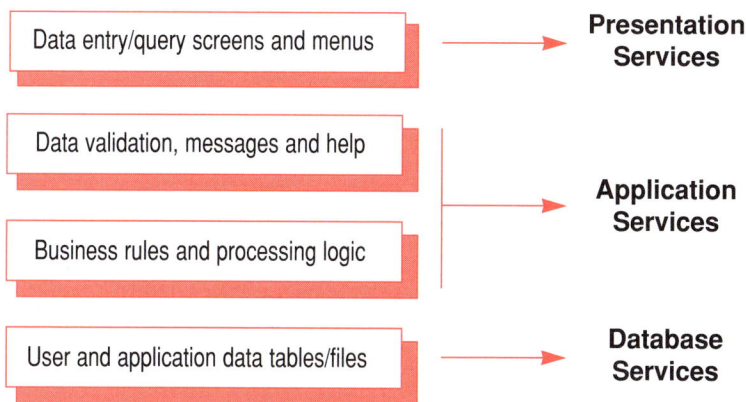

Data entry/query screens and menus	→	**Presentation Services**
Data validation, messages and help		
Business rules and processing logic	→	**Application Services**
User and application data tables/files	→	**Database Services**

The application services layer is where the real power of client/server application partitioning lies. If this middle layer can be isolated and run on separate computers, the client/server application can become almost infinitely scalable through the use of application or process servers, as I'll discuss later.

What Are Dumb, Thin, and Thick Clients?

How the application layers are distributed across clients and servers determines the client configurations you can use. Figure 1.3 shows the three most common types of client configurations. Most true client/server accounting applications offer deployment options using all of these configurations. The entire client/server application can also be deployed on a single device, such as a laptop computer or a standalone information kiosk. However, such deployments do not represent the mainstream of client/server accounting applications.

Dumb clients are those that handle only presentation services, leaving all application and database services to be deployed on the server. Dumb clients may include noncomputing devices such as touch-tone telephones

Figure 1.3: Three Common Client Configurations

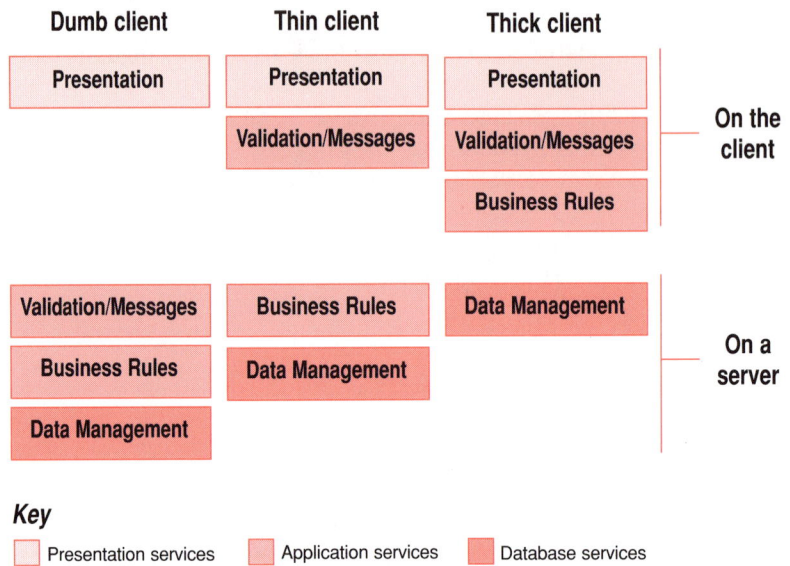

Dumb client	Thin client	Thick client	
Presentation	Presentation	Presentation	On the client
	Validation/Messages	Validation/Messages	
		Business Rules	
Validation/Messages	Business Rules	Data Management	On a server
Business Rules	Data Management		
Data Management			

Key

☐ Presentation services   ▨ Application services   ▧ Database services

or fax machines, on which no application client software actually resides. Internet browser software can also function as a dumb client because the application functionality being run in the browser is usually loaded from a Web server during a connection session and is not stored locally. Dumb clients usually interact with the accounting database server through a separate application server, such as a telephony or Web server.

Thin clients provide the application's presentation, validation, and error message services, with all of the main rule-driven application process services being located on remote application or database servers. The software used by the thin client may be stored locally or run remotely from an application or database server.

Thick clients provide presentation services and most or all application services locally. Because the client software is stored locally, one big disadvantage of the thick client deployment is that when the vendor updates the client software, the new release must be loaded onto every client — a maintenance headache compared to the simple one-time server-based upgrade required for dumb and thin client deployments.

The client configuration used determines the level of resources (CPU speed, RAM, and disk) required at the client (Figure 1.4) and whether the type of client can be varied to suit the business need. The fewer resources required, the less expensive the desktop equipment can be. Although this may not be an issue for single workgroup implementations of a client/server application, it can be a significant cost factor for enterprise-wide rollouts.

Figure 1.4: Client Resource Requirements

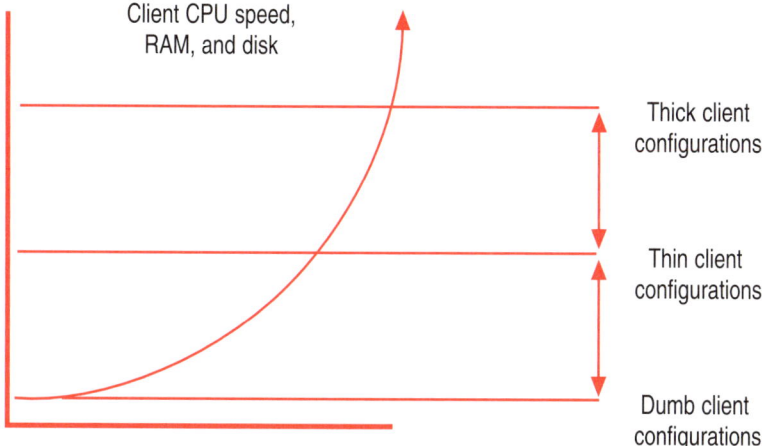

Enterprises implementing client/server accounting may connect a variety of client devices to their accounting applications, as illustrated in Figure 1.5. The wide range of possible clients emphasizes the client/server architecture's flexibility and ability to add value to business applications by integrating a variety of technologies and topologies.

Figure 1.5: Possible Accounting Clients

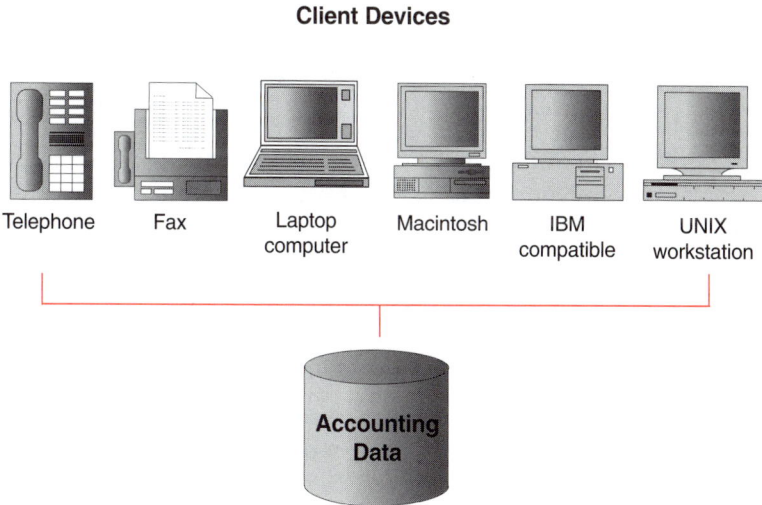

Client Devices

What Is the Role of the Server?

Most servers that can be used by a client/server accounting application fall into one of the five broad categories shown in Table 1.1. All client/server accounting applications use network and database servers, but the sophistication of the application partitioning determines whether the accounting application also supports the use of application, gateway, or process servers. Although these additional servers complicate accounting system deployment and maintenance , they also allow improvements in performance, scalability, and adaptability — benefits that can outweigh the drawbacks in some organizations, such as those experiencing rapid growth.

Table 1.1: Five Categories of Accounting Server

Server	Description
Network	Provides data file security and sharing, shared workgroup printing, communications, and electronic mail services
Database	Hosts the database and database management system for storing and managing the accounting data
Application	Provides services specific to an application, such as business rule and logic services, object and document management, decision support, and workflow services
Gateway	Links an accounting application to other network services, such as electronic mail, telephony, and Internet or intranet servers, and to accounting data on legacy systems
Process	Manages specific processes within an accounting application, such as financial reporting or batch transaction processing, including transaction posting and export

Choosing a single server operating system, such as a variant of Unix or Microsoft Windows NT, to power all servers helps provide a homogenous server environment and makes support and maintenance of client/server applications easier. But theoretically, a server may be based on a PC, a Unix machine, a midrange box such as an IBM AS/400, or even a mainframe (a handful of client/server accounting vendors now support the IBM DB2 mainframe database as a database server). This support of a wide range of server platforms connected across a network contributes to the flexibility of the client/server architecture.

When selecting a server, particularly a database server, for a sophisticated enterprise-level client/server accounting system, you should evaluate the server and its operating system against the criteria outlined in Table 1.2.

Table 1.2: Server and Operating System Evaluation Criteria

Criterion	Description
Operating system	The server should be compatible with popular client/server operating systems such as Unix, Microsoft Windows NT, IBM OS/2, and Novell NetWare to allow mixing and matching of operating systems.
Bus	The server and its operating system should support a 32-bit or 64-bit data bus and flat memory addressing, which allows use of large blocks of contiguous memory for optimum software performance.
Processor	The server and its operating system should support both single and multiple processors for performance scalability.
Disk	The server and its operating system should support multigigabyte disks or disk arrays for handling growth in transaction volumes.
Memory	The server and its operating system should support and be able to address at least a half gigabyte of RAM to allow the use of memory-based caches for high performance.
Adaptability	The server should employ a component-based physical architecture to allow "hot swapping" of disk drives, memory, and processor boards for easier maintenance.
Security	The server should support both software- and hardware-based security.
Availability	The server operating system should support processor clustering — the ability to use a group of CPUs as if they were a single processor — to ensure application availability despite an overload on any one server or processor.
Reliability	The server operating system should support disk mirroring, failover-clustering, or data replication to ensure nonstop, fail-safe computing if a server fails.

Servers are a crucial component of client/server systems. A poorly configured client impacts only one user, but a poorly configured server impacts all users who utilize its resources.

What Are Two-, Three-, and N-Tier Client/Server Applications?

An accounting application's granularity of design partly determines which clients and servers you can use with it. The granularity also determines whether the system can be characterized as two-, three-, or n-tier client/server. Most client/server accounting applications available today are two-tier or three-tier in architecture, but this will undoubtedly change as users come to see the potential benefits of n-tier architectures and as interest in Internet accounting matures.

Two-Tier Client/Server

A two-tier client/server application is run across clients and a database server, so the distribution of the application components is limited to the client or database server computers. Two-tier architectures are more restricted in scalability than are three-tier architectures. Scalability options are limited to upgrading the clients or the database server, even if only a single process, such as financial reporting, needs to be speeded up. Two-tier client/server systems tend to require highly configured clients, in the case of a thick client deployment, or an exceptionally highly configured database server, in the case of a dumb or thin client deployment.

Three-Tier Client/Server

Three-tier client/server accounting applications can use application servers between the clients and database servers. The middle tier helps reduce the need for resources on the client and database server and provides more scalability and easier connectivity to other technology services. By managing Internet connectivity, for example, an application server can prevent the client and database server from being burdened by connectivity responsibilities. Of course, the three-tier architecture requires more hardware and increases the overall complexity of the system and its management. But the additional tier also increases the range of deployment options by letting you put application services either near the client or near the database server, depending on whether the system is deployed on a LAN or a WAN.

N-Tier Client/Server

N-tier client/server essentially expands the middle processing tier, as shown in Figure 1.6. The n-tier scenario lets you run different types of application and process services, such as transaction posting or a specific financial report, on either clients or servers. N-tier applications may allow process scheduling as a way to provide flexible, demand-based process fulfillment.

Figure 1.6: Two-, Three-, and N-Tiered Architectures

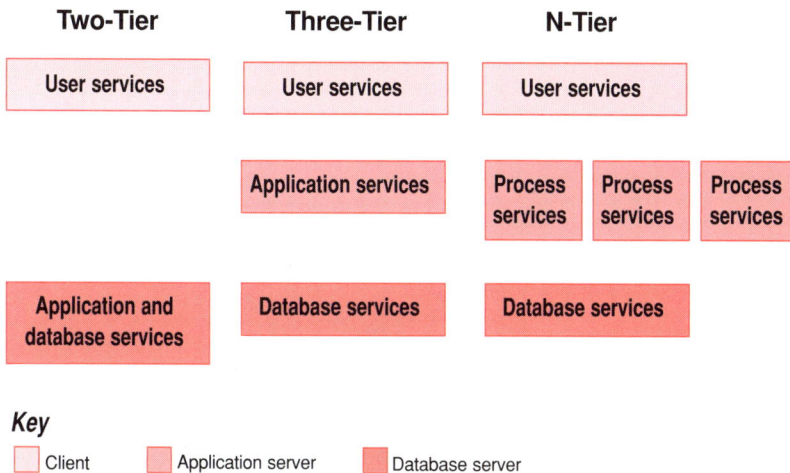

Two-Tier	Three-Tier	N-Tier
User services	User services	User services
	Application services	Process services / Process services / Process services
Application and database services	Database services	Database services

Key

☐ Client   ▨ Application server   ▦ Database server

An n-tier architecture may even let the application or the user choose the specific process server (or client) to run a particular process for the best performance. Figure 1.7 shows a screen for setting up process servers and allocating processes to them.

This screenshot shows the Process Server setup screen from the Dynamics C/S+ package from Great Plains Software. This screen is used to set up process servers for use by Dynamics and to allocate Dynamics processes, which may be local (client-based) or remote (server-based), to be run on those servers. A Dynamics process is one of a wide range of batch transaction and reporting processes from all modules of the system. The Dynamics process manager can automatically balance loads across process servers for optimum utilization of the server resources and can track the start and end times of process instances, such as report packs or batch processes, to assist in benchmarking application and server performance.

An n-tier client/server application offers by far the greatest scalability because any process can be allocated to its own dedicated resource for optimum performance. The system can be scaled on a process-by-process basis, if necessary. N-tier architectures use more server devices, generate more network traffic, and demand more sophisticated granularity in the design of the accounting application. In return, they offer the best possible scalability and performance options.

Figure 1.7: N-Tier Process-Server Maintenance Screen

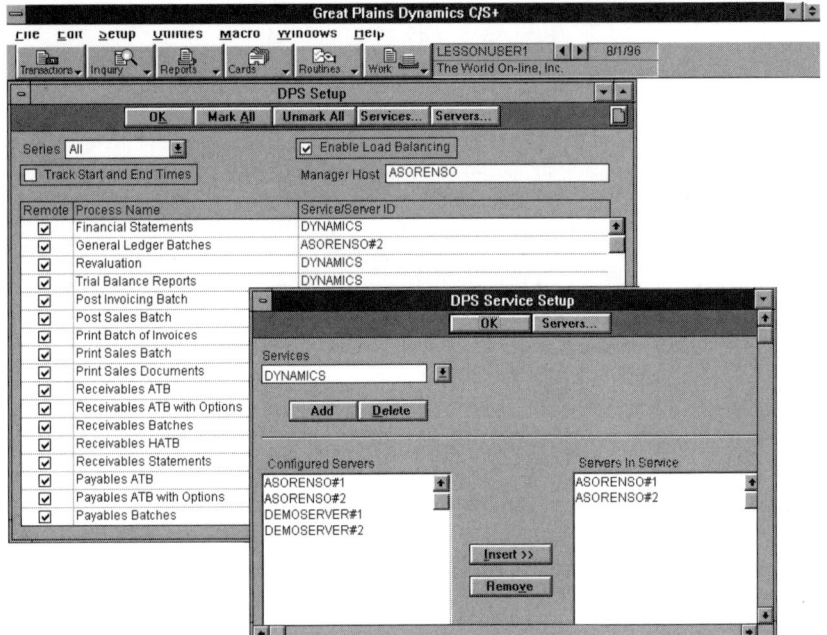

© Great Plains Software 1996

You should base your choice of architecture on the use of your accounting system and its projected future growth. For example:

If you expect	Choose
Stable, low-volume growth	Two-tier
Low reporting and batch processing needs	Two-tier
Minor integration of other technology (e.g., Internet)	Two-tier
Variable system deployment scenarios at different levels of business unit using LANs and WANs	Three-tier
Regular changes in business logic and rules	Three-tier
Extensive use of Internet or telephony integration	Three-tier
Variable-demand batch processing	N-tier
Variable-demand report processing	N-tier
Multiple feeder systems	N-tier

Conclusion

With client/server increasingly incorporating Internet and intranet clients and servers, there is little doubt that the architecture has a long life ahead of it. Three-tier and n-tier client/server accounting applications stand the best chance of adapting to new infrastructures, such as the Internet, because the middle tier can be used as a means for accounting clients and database servers to communicate with other application servers, such as a Web or intranet server. As accounting applications become more granular and are eventually deployed as collections of functional components, the process servers in the middle tier will become more important in the overall management of what will be the ultimate distributed computing architecture. However, it remains a challenge to set up, maintain, and monitor three- and n-tier client/server applications, to manage interapplication connectivity over a network of distributed resources, and to maintain network throughput in a world of local, wide-area, and Internet connections.

This challenge will only increase as accounting applications are redesigned as individual, collaborating functional components. In the process, two primary benefits of client/server — flexible deployment and scalability — will come under pressure as it becomes more difficult to maintain communications between objects used by distributed applications and to maintain acceptable levels of application performance across the busy network bandwidth. In this scenario, recentralization of applications is likely, in combination with more use of offline client applets (small, self-contained applications that accomplish some discrete part of the accounting function) that store their data locally and forward it to the central servers only through periodic online conversations.

QUESTIONS FOR VENDORS

1. Has the package been specifically written or rewritten for the client/server platform?

2. When was the first live implementation of the client/server package?

3. Which client/server platform (server, database, and GUI) did the vendor actually develop the package on?

4. Which client/server platform is the package ported to first when new versions are released?

5. How many live installations (as opposed to legacy systems) are there of the client/server package?

6. How many live installations of the client/server package exist on the platform you plan to use?

7. Does the package deploy as a two-tier, three-tier, or n-tier application?

8. What are the processor, memory, and disk recommendations for the application on the client and on the application and database servers?

9. If the package is an n-tier application, can users allocate processes to specific process servers and schedule or reschedule the processes?

10. Can the package support configurations that combine dumb, thin, and thick clients?

Chapter 2

Relational Accounting

Relational Accounting 101

Relational accounting is the use of a relational database management system (RDBMS) to store and manage accounting data. Over the past decade, the RDBMS has become the technology of choice for managing data generated by online transaction processing (OLTP) — the high-volume processing of predictable, structured transactions such as ATM withdrawals, airline reservations, and catalog orders. Along the way, Structured Query Language (SQL) has become the standard means (an ANSI standard, in fact) of accessing data in an RDBMS. SQL can be used either in its native form or through SQL-based data-access application programming interfaces (APIs) such as Microsoft's Open Database Connectivity (ODBC). The RDBMS and SQL emerged from research at IBM labs during the 1970s and from the pioneering work of Dr. E.F. Codd, known as the father of the relational model. The commercial realization of this research was a database architecture that is both adaptable to change and easy to access, solving two important problems of traditional mainframe network and hierarchical databases and of other positional file management systems.

Instead of the more traditional single file with multiple record types, the relational database model demands a more atomic organization in which the data is broken down into smaller, discrete structures logically represented to users as tables containing rows and columns. To change the database, you add new tables to it, new columns to existing tables, or new relationships between tables. To view the data differently, you use joins. A join lets you build a new logical view of the data by combining two or more tables that have at least one column — or join key — in common. Figure 2.1. illustrates two relational database tables and a new logical view of those tables obtained by using a join. SQL is the language used to create and manage the database, tables, columns, and joins and to query the data.

Figure 2.1: Joining Two Relational Database Tables

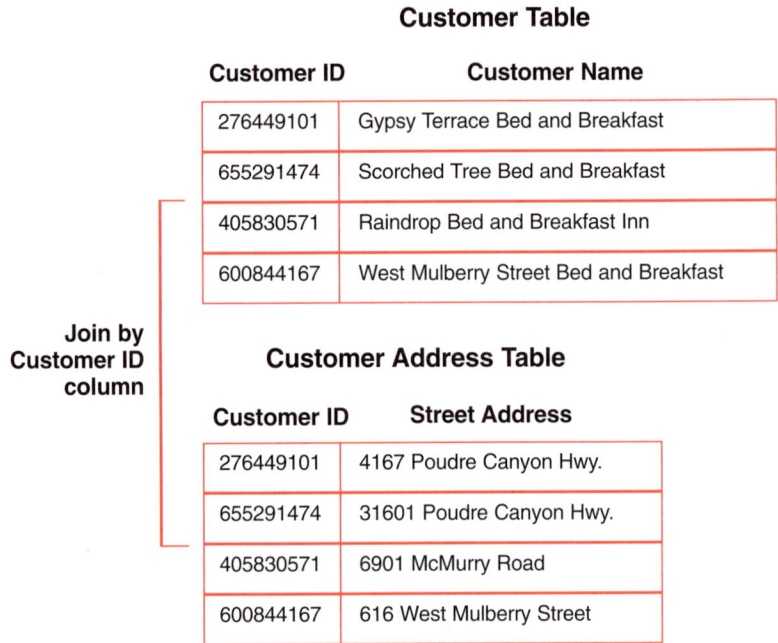

Customer Table

Customer ID	Customer Name
276449101	Gypsy Terrace Bed and Breakfast
655291474	Scorched Tree Bed and Breakfast
405830571	Raindrop Bed and Breakfast Inn
600844167	West Mulberry Street Bed and Breakfast

Join by Customer ID column

Customer Address Table

Customer ID	Street Address
276449101	4167 Poudre Canyon Hwy.
655291474	31601 Poudre Canyon Hwy.
405830571	6901 McMurry Road
600844167	616 West Mulberry Street

New Joined View of the Two Tables

276449101	Gypsy Terrace Bed and Breakfast	4167 Poudre Canyon Hwy.
655291474	Scorched Tree Bed and Breakfast	31601 Poudre Canyon Hwy.
405830571	Raindrop Bed and Breakfast Inn	6901 McMurry Road
600844167	West Mulberry Street Bed and Breakfast	616 West Mulberry Street

Relational accounting has some drawbacks, however. In most cases you must buy the RDBMS licenses you need separately from your accounting application licenses. In the past, most accounting applications came with a built-in proprietary database or file system that had either been licensed from a third party or supplied by the accounting software vendor. The cost of the data management system was therefore bundled into the price of the application modules. Built-in data management systems required little or

EVALUATION HEADS-UP

- All client/server accounting systems use relational databases to manage their accounting data. But all relational databases are not the same in terms of functionality, resource overhead, scalability, or multiplatform support. If you don't have a relational database in place, you should carefully determine which RDBMS is right for your information management needs before you buy the accounting application that runs on top of it.

- When evaluating a relational database, check whether it supports rollback and roll-forward transaction recovery, stored procedures, and triggers and whether it can deliver more advanced functionality, such as parallel processing or replication. Rollback and roll-forward transaction recovery saves time when recovering from system failures and helps maintain the integrity of your accounting data. Stored procedures and triggers can improve performance and increase functionality. Parallel processing and replication, while not yet used to full advantage, also hold great promise for future performance and functional improvements.

- Because relational databases are notoriously resource-hungry, the configuration of the database server (in terms of bus, CPUs, disk, and memory) is critical to getting optimum performance from client/server accounting applications. Having a lot of memory on the database server can dramatically improve overall system performance, so budget for as much memory as you can afford.

- Applications that support more than one RDBMS do so by using embedded SQL or middleware, such as Microsoft's ODBC, to connect to the database. Although this approach offers database choice and expands the market for the package, it may not provide the best performance on every supported database or take full advantage of database-specific functions. Applications that use database-specific stored procedures usually support only one or two databases, but such applications are likely to run faster, reduce overall network traffic, and be better adapted to native database functionality. Unless you need a multidatabase

Continued

deployment for your accounting systems, you should choose an application that uses stored procedures over one that does not.

- A client/server accounting system always needs a database server to manage the relational database, but it may also need additional servers to run other database architectures, such as multidimensional or object databases. This is likely to be the case if you intend to use OLAP or image-management technology in conjunction with your accounting system, for example. If you plan to use these complementary technologies, you must budget for acquiring and maintaining the additional database servers.

no administrative overhead to maintain and tune. But most RDBMS systems require a full- or part-time database administrator because extracting the best performance from a sophisticated RDBMS requires regular monitoring, maintenance, and tuning.

Because of the overhead incurred in delivering some of the benefits of relational accounting, an RDBMS may be slower than other data management options, especially when posting new transactions or updating existing transactions on low-end equipment. Also, RDBMSs are typically resource-hungry. Most RDBMS products deliver their best performance on a multiprocessing database server configured with a lot of spare memory and extra disk capacity. Consequently, the server configurations purchased for an RDBMS-based accounting system tend to be more expensive than for alternative data management options. You can also expect to purchase at least one separate computer for the sole purpose of running an RDBMS.

Despite these drawbacks, accounting software vendors believe that relational accounting delivers more advantages than disadvantages, especially for businesses that expect rapid growth or require better access to their accounting data than they've been able to achieve with a traditional database. One of the biggest points in favor of relational accounting is simply that it puts accounting vendors squarely in the accounting application business and takes them out of the database business. By passing the responsibility for data management to a third-party vendor focused on building database software, accounting vendors can allocate more R&D resources to their applications instead of to maintaining proprietary databases or unique file management systems.

By riding the database vendor's R&D investment, accounting vendors should be able to build richer functionality in their accounting applications and leverage advantage from the RDBMS vendor's enhancements to its database product. A good example of such leveraging is the case of replication functionality. All leading RDBMS products now offer some form of data replication for automatically copying data from one database to another, for example for use in intercompany processing or consolidation reporting. Accounting-application vendors can take advantage of this database feature without having to spend the time and money to develop replication functionality themselves.

Every RDBMS uses some version of SQL to create, maintain, and query the database. Despite the known deficiencies of SQL for producing complex matrix-based financial reporting and the variety of commercial implementations of the ANSI SQL standard, the fact remains that the applications software market has adopted SQL as its lingua franca for accessing data in an RDBMS. Consequently, transaction data stored in relational accounting systems can now be directly accessed by a wide range of SQL-compatible desktop worksheet, query, and reporting tools. More than anything else, the commercial acceptance of SQL as a common database access language has encouraged vendors to develop a whole new generation of powerful decision support tools. Instead of limiting your choice of decision support tools to those provided by the accounting vendor, you can now pick and choose from a range of inexpensive desktop packages that enhance the value of your accounting information asset.

By standardizing on SQL as their database management language, RDBMS vendors have also done users of relational accounting applications the favor of giving them more options. In the past, most accounting systems offered just one option for their data management system, namely the vendor-supplied database or file system — take it or leave it. Now many vendors have isolated the database access layer from the rest of the application code, so their applications can be connected to any one of several popular RDBMS products. Clearly this offers accounting system managers more choice of database engines, but more importantly it allows a business two further advantages. First, an enterprise may use the same applications worldwide but use different RDBMS engines to manage the data in different locations for reasons such as performance, cost, or local support. Second, an enterprise can choose to change its RDBMS — for example to benefit from a new technology or the better price/performance of a competitive product — without having to undergo the costly process of changing the entire accounting package.

Using a relational database also encourages application developers to remove business rules, logic, and security from the application code and store it in the database as stored-procedure or trigger objects. From a developer's perspective, the benefit is that the business rules are centralized in the database, rather than being duplicated and distributed over all the desktop clients, and are thus easier to maintain. From a system management perspective, the benefit is that any application (e.g., a budgeting tool, even if it's a desktop tool rather than part of the accounting application) that attempts to access the accounting data is subject to the same security and constraints as a bona fide accounting application user.

Furthermore, the cost of relational accounting is coming down. Recent benchmarks released by Microsoft Corporation for its SQL Server 6.5 RDBMS running on a DEC Alpha server under the Microsoft NT Server operating system suggest that exceptional performance can be expected from relatively low-cost computer platforms. Currently, the starting cost for a five-user Microsoft NT Server operating system and SQL Server RDBMS combination is under $2,000, and you can buy sophisticated RDBMS products for use by workgroups of up to 20 users for less than $1,000. Those who need relational accounting on a mainframe will find even that option provided by client/server accounting vendors who support IBM's mainframe DB/2 RDBMS.

Although object and multidimensional databases will play an increasingly important role in client/server accounting systems in the future, relational accounting is the name of the game for at least the rest of the 1990s. By converting to relational accounting, you no longer take a risk, but simply recognize that database choice, scalability, transaction integrity, centralized business rules, and other benefits provide a solid foundation for a business management system.

Why Should I Care About Database Access?

The way a client/server accounting application accesses the data in the database can have a significant impact on system performance. Vendors use a variety of approaches that essentially depend on whether they want their accounting applications to support more than one RDBMS. Support for more than one RDBMS gives users a choice of database engine and gives the vendor a bigger potential market for its product, but it makes the application itself more tricky to develop and maintain because every RDBMS has a different feature set that can be leveraged (or not) by the application. Also, despite the SQL standard, every commercial RDBMS uses slightly different and extended versions of SQL, so a given SQL statement may not access data with the same results (or any result) across different RDBMS products.

Embedded SQL and ODBC

Vendors who want their accounting applications to work with multiple databases may embed native SQL statements in their application source code and be forced to maintain multiple versions of the embedded SQL, one for each supported database. Alternatively, they can write their database access code to use a multidatabase API or middleware program, such as Microsoft's ODBC, to provide access to multiple databases via a single access "layer" (Figure 2.2 illustrates this approach). In its initial versions,

ODBC was criticized for being a lowest-common-denominator approach that may not be as fast as other methods nor support as rich a set of database access functions as does the native SQL syntax for the database or a specialized driver built to support access to a specific database. However, version 3.0 of ODBC has largely overcome these deficiencies.

Figure 2.2: Multidatabase Access Via Microsoft ODBC

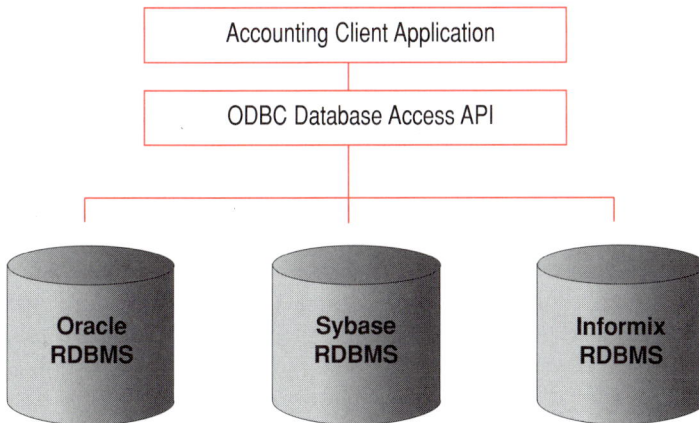

Multidatabase Access Via Microsoft ODBC

In addition to connecting to a choice of RDBMS engines, applications that use embedded SQL or ODBC-based access can connect more than one database to individual modules within the application. For example, a client/server accounting system's financial modules may run on an RDBMS from Sybase Corporation while its distribution modules run on a database from Oracle Corporation. Such flexibility may deliver a performance or functional benefit. But even with recent improvements to ODBC, accounting applications that access databases through embedded SQL or ODBC-type middleware may be less efficient or slower than those that use specialized native database-specific drivers or, better still, execute server-based stored procedures.

Stored Procedures

Stored procedures are pieces of SQL and conditional logic code stored and maintained in the database as discrete objects. Client applications can call a stored procedure from the database and execute it on the database server to carry out a specific task such as inserting, updating, or deleting a transaction or running a complex process such as posting orders, calculating payroll, or printing a batch of reports. Stored procedures are very efficient because when the code is created and stored, the database manager reviews, tests, and optimizes it to ensure it's correct and is executed as efficiently as possible. Subsequently, the database manager itself runs the procedure, which ensures that the code can take full advantage of the available database server resources. Stored procedures centralize much of the application business rules and logic in the database instead of dispersing it in the application code across many application clients. This repository approach makes it easier for vendors and accounting system managers to maintain and improve the rules and logic on a procedure-by-procedure basis.

Stored procedures are network-efficient because the client application is required to ship very little information to the database server to run the procedure. Instead of sending a packet containing a complex, embedded SQL statement — which may be dozens or even hundreds of lines of code — from client to server, the application can send a compact instruction known as a remote procedure call that simply passes a few parameters and requests the database manager to run a specific procedure using those parameters. The small remote procedure call helps keep network traffic low and ensures better overall network performance. An accounting application that uses stored procedures is likely to be more efficient and faster, generate less network traffic, and be more closely optimized for use with a specific database than an application that does not use stored procedures.

The use of stored procedures also indicates that the vendor has isolated core business rules and logic by removing them from the application code. Business rules and logic that have been isolated as stored procedures may eventually be removed from the database and deployed as discrete functional objects on their own server. This is one of the first steps toward building more flexible and granular accounting applications and is a move toward a system based on cooperating software components known as business objects (for a discussion of component accounting and business objects, see Chapter 10). However, one drawback of stored procedures is that they are database-specific and not easily portable to other databases. For this reason, accounting applications that use stored procedures often limit themselves to supporting only one primary RDBMS line.

How Does Relational Accounting Protect My Accounting Data?

Maintaining the integrity of accounting data is a concern that is foremost in the minds of all accounting system managers. One reason an RDBMS is usually slower at inserting, updating, and deleting data than is a basic file manager is because it uses a transaction log and maintains referential integrity.

Transaction Logs

All leading RDBMS systems offer transaction logging, which means the database manager automatically populates a log with data about the activity in the database. The log is a useful administrative tool for analyzing and tuning database performance. But maintaining the log adds time to every transaction — time that can add up in high-volume transaction-processing environments. Logs also take up space, demanding more disk resources on the database server, and must themselves be maintained and monitored for efficiency. Most file systems don't maintain transaction logs and consequently don't suffer the associated performance hit.

The primary benefit of the log is that the database manager can use it to recover from system failures. All popular RDBMS products are quite reliable, but the applications that connect to them and the networks they run on may not be, so system crashes are a fact of life. After a crash, the database must be restarted. As part of the startup process, the database manager checks the log to determine whether any transactions sent to the database were not completed. Any incomplete transactions are rolled back — effectively undoing the impact of the transaction— so that the database is restarted in a consistent state. All leading commercial RDBMS products support transaction rollback. Some also support roll-forward recovery, which replays the logged activities from the point at which the database was in a consistent state up to the point at which the system failure occurred. Roll-forward reduces the rekeying that would be necessary if the failure occurred in the middle of posting a batch of accounting transactions. Rollback and roll-forward recovery through transaction logging are two benefits of relational accounting that ensure more reliable accounting systems.

The integrity of the data in an RDBMS is also protected through the use of referential integrity, and column and table constraints, that enforce database "business rules." Referential integrity is used to create relationships between tables to prevent data from being entered into a column in one table unless the data exists in a column in another table; for example, to ensure that only valid account codes are posted to a journal table. Referential integrity may also be used to prevent table columns from being updated or columns and rows in tables being deleted if the data is stored in other, linked tables. For example, this prevents an account code from being deleted from an account table if it is stored as part of a transaction in a journal table. Column and table constraints link business rules to columns or tables to prevent the entry of incorrect data. For example, a column constraint could ensure only valid business dates are entered into a transaction date column, and a table constraint might be used to prevent entry of new rows into an order table if a customer has exceeded a credit limit. Referential integrity, and column and table constraints, can also be implemented through the use of triggers — another functional benefit of relational accounting.

Table Triggers

Most popular RDBMS products support table triggers. A trigger is a piece of code that's stored in the database and associated with a particular table (or sometimes a column within a table). The trigger is automatically "fired" (i.e., the code is executed) whenever a specified database event, such as an insertion or update to the table, takes place. Triggers are often used to maintain referential integrity because they effectively intercept data before it reaches the database, and so can be used to apply business rules to the data before passing it to the database manager.

More than one trigger may be associated with a table. Consider a general ledger journal table, for example. An insert trigger can be used to fire a stored procedure that checks that the incoming journal balances and all the codes and values contained in the journal are valid against other table data. If the data integrity checks fail, the journal will be rejected, preventing an invalid transaction from reaching the ledger. An update trigger can preserve the audit trail by allowing certain information on the posted journal — a description, for example — to be changed, but not other information, such as the journal ID or date. A delete trigger can prevent any posted journal from being deleted from the ledger table, thus ensuring posted transactions are never accidentally deleted.

Referential integrity is especially critical when you consider that accounting data in an RDBMS is potentially open to access and manipulation from other desktop tools that know nothing about accounting rules. By placing logic in the database using triggers, you can make sure all applications accessing the accounting tables are subject to the same integrity checking. Triggers can thus be regarded as playing the role of data guardian, calling efficient stored procedures to do the actual work of checking the data against specific business rules.

Triggers are widely useful in a client/server accounting system even beyond ensuring referential integrity. For example, they can

- automatically generate and store in a separate table audit records that log all activity on sensitive accounting tables such as the chart of accounts, check payment tables, and payroll tables

- send an e-mail alert or information message to a user or group of users when a certain threshold is reached or exceeded in the table data, such as a payment of more than $1 million or an order for a specific product

- initiate stored procedures to carry out a wide range of accounting processes — for example, generating month-end reports when an update to a period table indicates that the current period is closed

Triggers can effectively deliver many benefits to any application that interacts with the accounting database. Consequently, accounting applications that don't support triggers are probably not designed to take full advantage of the features of a specific RDBMS. On the other hand, overuse of triggers or the use of highly complex triggers can slow database performance, and triggers may be difficult to debug because their processing takes place under the covers, rather than by explicit application code.

What Future Benefits Can I Expect from Relational Accounting?

Most popular RDBMS products now support some level of parallel processing and replication functionality. As these technologies mature, accounting applications will likely take advantage of their potential to add functional value.

Parallel processing is the use of multiprocessing computers to execute database tasks in parallel rather than sequentially. For example, parallel processing might let the database manager concurrently import accounting transaction feeds, query account balances, and run a financial statement. It's easy to see how parallel processing can lead to dramatic performance improvements at times of peak processing loads such as period-end closings or consolidation reporting workflows.

Accounting applications can also leverage database replication functionality. Replication allows data to be copied automatically to synchronize two or more physically separate databases. Replication can be used to

- publish reports and other accounting data to subscribing servers for dissemination to users via e-mail or the Internet

- upload retail outlet or business-unit data for the day, week, or month to regional or headquarters databases for consolidation purposes

- download specific transactions from regional or headquarters sites to selected business units for automating intercompany or allocation processing

- transfer aggregated data from transaction accounting systems to decision-support data warehouse or datamart systems

Few client/server accounting applications can currently take any real advantage of parallel processing or replication features in the databases they support. But both technologies have the potential to deliver performance and functional benefits and are certain to be areas in which accounting vendors will focus some of their R&D effort in the near future.

What About Btrieve and Other File Managers?

Today, many accounting systems used by small workgroups of up to 50 users run on local area networks using Novell Corporation's NetWare, Pervasive Inc.'s Btrieve, or some other indexed sequential access method (ISAM) file management system. Despite the general transition to relational accounting, file managers do have some benefits. Because they are relatively simple structures, they offer fast performance at sites with low transaction volumes. They have relatively low resource overhead, so they are ideal for use on low-end computers such as laptops. They are usually bundled with the accounting software, so there is no need to go out and buy another database engine. And file managers are low-maintenance — except when they go wrong — which means you may not need to hire a database administrator.

File managers have become more sophisticated over the years and now include more options for tuning the file system, maintaining file integrity, and accessing data through SQL. However, when compared to today's leading RDBMS products, file managers suffer from a number of drawbacks:

- They're missing sophisticated functionality such as replication, parallel processing, and the ability to handle a variety of data types other than text and numbers.

- They lack the scalability and portability that allow database engines to accommodate very high transaction volumes and run on a range of powerful servers.

- They don't provide the ease of integration demanded from accounting systems in a world of relational data designed to be accessed by SQL.

In short, there is nothing wrong with Btrieve or other file managers when used for workgroup-level accounting systems where transaction volume is low, scalability needs are modest, or business accounting demands are functionally simple. But for organizations that don't fit this profile, the benefits of file managers do not outweigh their drawbacks. Furthermore, many larger organizations store much of their operational data in relational formats, so it just does not make sense for accounting to use something different that makes it more difficult to integrate applications.

What About Multidimensional Databases?

The RDBMS may be the king of OLTP, but many analysts and users believe multidimensional databases are superior for decision support and online analytical processing (OLAP). OLAP is a marketing term used to describe the complex, user-driven, ad-hoc queries managers and analysts need to be able to do to manage a dynamic business environment and assist with decision support.

OLAP applications typically use a multidimensional database management system (MDBMS) to provide the speed and data navigation capabilities such decision support activity demands. The key difference between multidimensional and other database technologies is that multidimensional technology can provide a data value for every intersection of the information dimensions managed by the MDBMS. For example, if you track actual sales dollars, you may want to analyze those actuals by six dimensions, each comprising n elements: line of business (let's say 3 elements), product SKU (1,000 elements), city (50 elements), state (10 elements), salesperson (20 elements), and month (12 elements). A multidimensional database would be able to find a dollar value for every intersection of every element across all six dimensions — potentially some 360 million or so values in any one year. Figure 2.3 illustrates the "star" nature of a multidimensional database by comparison with conventional two- and three-dimensional information matrixes.

Figure 2.3: The Multidimensional "Star"

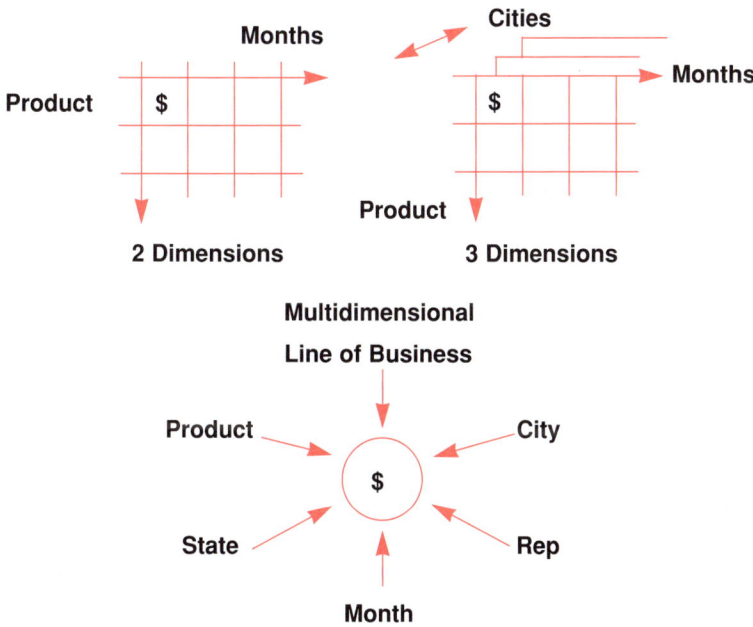

The Multidimensional "Star"
Through the use of sparse matrix technology, the MDBMS may not actually store a value if one doesn't exist for a particular intersection, thus reducing the size of the database and speeding up data retrieval and calculation.

Some MDBMS products precalculate all intersection values, and others calculate them on the fly; regardless, the MDBMS is a very efficient way to access a precise data value across multiple analysis dimensions compared to the complex SQL queries necessary to achieve the same end with a relational database.

Because of the MDBMS structure and the fact that small multidimensional databases can be managed entirely in memory, you can expect very fast performance during ad-hoc analysis. Multidimensional structures provide an optimum architecture for data drilldown, focusing, and rotation. Drilldown means navigating from summary accounting data to detail transaction data. Focusing means expanding and collapsing data — from year to quarter to period, for example. Rotation means pivoting the dimensional axis of a matrix — for example, switching from product sales by month to product sales by city.

Some IS departments are considering or implementing a client/server accounting architecture that encompasses a relational database server for transaction processing and a multidimensional database server for decision support and financial reporting. Despite the cost and administrative burden of this approach, a dual database server architecture can ensure that decision support activity receives as much emphasis as transaction processing. Interest in OLAP has stimulated accounting vendors to improve their internal decision support functionality and to supplement it with links to third-party OLAP query tools based on multidimensional database engines. Now that Microsoft Corporation has entered the multidimensional database market with its own server products, OLAP is soon likely to become a part of every accounting department's standard toolset.

How Can an Accounting System Use Object Databases?

Although the relational architecture dominates the client/server accounting software market for transaction management and multidimensional databases continue to gain popularity for decision support, object database management systems (ODBMS) are likely to become increasingly important as the range of types of data managed becomes more extensive. The objects an ODBMS manages are discrete pieces of software that encompass both data and the behavioral methods used to access and manipulate that data. Many document imaging systems, for example, use an ODBMS to store and manipulate digital versions of documents, such as invoices and checks. The database stores not only the information on the document (e.g., the data for a sales invoice), but also the code that determines how that information can be manipulated (its methods). Objects are usually represented as a kernel of data surrounded by the manipulation methods (Figure 2.4). Applications that want to use an object can access its data only through the prescribed methods that define the object's behavior. The object structure provides a clear interface for integrating the object with other applications and ensures the integrity of the object by preventing access to its data except through its methods.

An Invoice Object

It's reasonable to assume that in the future, accounting systems will include the capability to manage not only images and documents, but diagrams, sound, and video as well. In fact, the accounting systems of the future will themselves almost certainly be assembled on the fly from software components called business objects that will be stored either locally on a client or on a local, remote, or Internet/intranet server. The database technology most likely to be used to manage these objects will be an ODBMS.

Figure 2.4: An Invoice Object

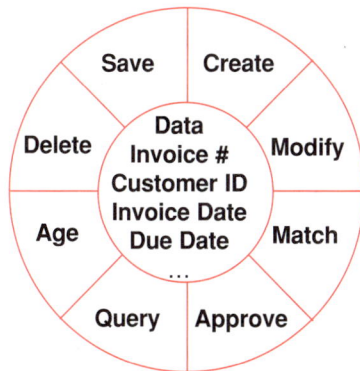

Conclusion

The RDBMS is currently the foundation for all client/server accounting applications, but multidimensional and object databases will become increasingly important parts of the accounting architecture. Multidimensional databases are likely to become the preferred database structures for financial reporting and decision support, leading to wider use of a dual-processing accounting architecture: a relational system for transaction processing and a multidimensional system for reporting. Object databases will become popular for storing documents and images associated with accounting transactions, but more importantly will become the system of choice for storing and managing objects used by workflow management software. Alternatively, so-called universal database servers, capable of managing relational, multidimensional, and multimedia data, could become the preferred foundation for accounting systems that manage increasingly complex processes and a wider range of data than simply text and numbers.

QUESTIONS FOR VENDORS

1. Does the application include a database engine, or do you need to have a specific database engine in place to run the application?

2. Can the application be deployed against more than one database engine?

3. Can the application be deployed using the database engine on any of its supported server platforms?

4. Does the application support the latest version of the database engine or only a previous version?

5. Does the application communicate with the database using embedded SQL, "native" database connectivity drivers, or middleware such as Microsoft's ODBC?

6. Does the application use stored procedures or triggers on the database server?

7. Can the application take advantage of specific database functionality such as replication or parallel processing?

8. Does the application require the use of more than one type of database engine, such as relational and multidimensional, and must the additional engine be purchased separately?

9. What is the CPU, disk, and memory configuration recommended for use on the database server?

Chapter 3

Graphical Accounting

Graphical Accounting 101

Graphical accounting is the use of a graphical user interface to present the accounting system functions and data to users. It's taken a few years, but now practically every major accounting vendor has released a true graphical user interface (GUI) for its accounting applications. Of course, graphical accounting in some shape or form has been around for some time. Accounting software for the Apple Macintosh has always offered a GUI, and SOHO (small office, home office) packages such as Intuit's Quicken and Peachtree Accounting for Windows were early to market with GUI versions of their products for PCs running the Microsoft Windows GUI.

For many users of legacy accounting systems on minicomputers and mainframes, "graphical accounting" equates to a color choice for the displays on their terminals: either green on black or amber on black. However, true graphical accounting makes much better use of color and incorporates graphical interface symbols, sometimes called "widgets," to make the user interface more intuitive, easier to learn and use, and more interesting.

One of the most important benefits of graphical accounting is the homogenization of accounting user interfaces. In the past, accounting vendors built their own interfaces, chewing up resources that could have been spent on functionality and resulting in a profusion of interface designs ranging from difficult to use to bizarre. Now, by and large, user interfaces for accounting applications look very much like those of other popular desktop applications —it can be hard to tell the difference between the menus and toolbars of accounting packages and those of desktop applications such as Microsoft Word or Excel. This similarity makes accounting systems easier to learn and helps users become more productive faster. Vendors also benefit because they can concentrate on improving and expanding functionality to increase the breadth and depth of their applications and let deep-pocket companies such as Microsoft and Apple concentrate on interface design. The fact that every vendor is using essentially the same standard interface has another advantage: it's become easier for users to compare packages and spot good, bad, and innovative interface designs.

If you were to break down accounting systems into three main functional areas, they could be

- building structures
- transaction entry and processing
- query and reporting

Graphical accounting delivers real benefits in all these areas.

Accounting systems are full of hierarchical structures: charts of accounts, organization reporting trees, sales territories, bills of material. Graphical accounting provides the browsers that let you efficiently maintain and navigate these hierarchies by expanding and collapsing the nodes in the data trees. Many accounting structures also consist of logically linked categories of information — for example, a customer card may link customer account information, address, terms, and contacts. Structures of this type can be represented graphically very effectively using a visual tab folder to view and maintain the different but linked customer information items. Hierarchy browsers and tab folders can also be very effective when used together, as Figure 3.1 shows.

ScreenHelp: This screen shows the General Ledger maintenance tab folder for an accounting application built using Powersoft's PowerBuilder. The browser on the left lets the user navigate the chart of accounts by collapsing and expanding nodes in the account hierarchy. The form on the right is used to insert new accounts and view and maintain existing accounts. A navigation bar of VCR buttons above the form lets the user navigate the chart of accounts and find accounts using the account code as a lookup.

Not all accounting transactions are simple, repetitive inputs; many are based on paper documents such as invoices, requisitions, and orders. These documents have header, detail line item, and totaling areas that can be represented graphically and very effectively as WYSIWYG (what you see is what you get) forms that users fill in on screen. Such on-screen forms can also be immediately output on plain paper and mailed or distributed electronically without the need for special, costly preprinted forms.

Although there are faster ways to input transactions than through a GUI, the GUI makes it much harder for users to make simple input errors by providing a variety of controls for restricting input choices and letting users select items from lookup list boxes. Increasingly we are moving

Figure 3.1: A Combined Browser/Tab Folder Interface for
Maintaining General Ledger Data

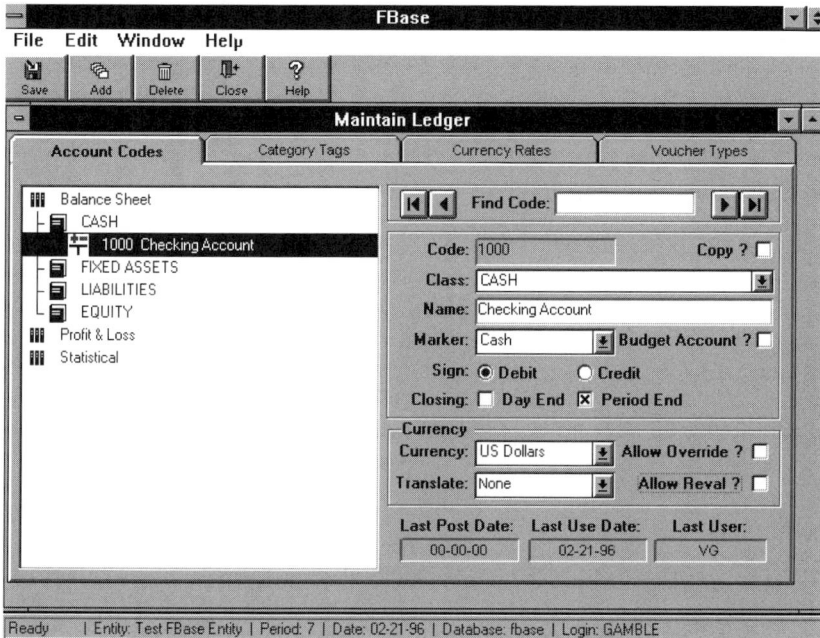

© FBase Inc. 1996

toward a world in which transactions will reach accounting systems through
electronic commerce, so transaction entry may soon become a dying art.

Querying, reporting, navigating, and investigating data is where graphical accounting really shines. Query-by-form (also called query-by-example)
provides an intuitive way to build complex queries — you simply enter
query conditions and operands in the same boxes you use to enter data. You
can manipulate the query results in many ways. You can use balance queries
to drill down, from summary through detail information all the way to the
original transaction, using just the mouse to double-click down and up the
drilldown navigation path. You can scroll the grid displaying the results of
the query vertically and horizontally and can resize, re-order, and re-sort
columns on the fly. You can highlight ranges of data with the mouse, and
with a click on a toolbar button, display data as a chart, export it directly
into a worksheet, or cut and paste it into a word processor. And you can use
color to great effect to highlight threshold-based exceptions or indicate differences such as positive and negative numbers, making it easier to assimilate information.

EVALUATION HEADS-UP

- Although a few client/server accounting applications support character-based screens, almost all now expect users to utilize one or more graphical user interfaces (GUIs), with Microsoft's Windows 3.1x and Windows 95 GUIs being the most commonly supported. Fewer vendors every year support other GUIs, such as those that form part of the Apple MacOS or the Open System Group's Motif GUI.

- There is no reason to accept a poor GUI design. Accounting users will have to live with the GUI day-in and day-out for years. A poorly designed GUI saps productivity and increases user frustration with a package. Evaluate the GUI with the same degree of focus as you do other accounting application functionality.

- Accounting applications that offer usability functions similar to Microsoft Windows cue cards, wizards, tooltips, and context-sensitive help systems are easier to implement and use, resulting in faster implementation, lower cost, shorter learning curves for new users, and fewer operational errors.

- Using tab folders to organize linked data and hierarchical browsers to navigate parent-child structures can make creation and maintenance of complex accounting structures easier and the finished product more intuitive. Support for drag-and-drop capabilities helps maintain visual relationships between data. WYSIWYG data entry and query forms help users more easily make the transition from paper-based forms to electronic transactions, while graphical report builders can significantly simplify the business of building and visualizing complex multilevel reports.

- Support for a multidocument interface ensures that graphical accounting systems let users view and use multiple screens simultaneously. Such multitasking capability helps ensure higher productivity and lets users avoid having to close single-screen forms before they can move around the system.

Continued

EVALUATION HEADS-UP continued

- Look for extensive keyboard options in the application menus and for shortcut or accelerator key combinations. Proficient keyboard users, such as accounting clerks, can be much more effective when they don't have to use the mouse for every interaction with the application.

- For users focused on rapid, heads-down data entry, look for systems that support character terminals or provide special non-graphical transaction-entry screens to facilitate fast input of high volumes of data.

Graphical report building provides a visual way to construct a report and see how it will look when printed — both at the same time. As you build the report, you can populate it with sample data to check whether the data is correctly formatted for the audience. Before printing the report, you can preview it graphically on screen to check alignment, layout, and page breaks. Alternatively, you may not print the report at all, choosing instead to save and view it on screen or to send it as an attachment to an e-mail message to people who need to use the report data. Although complex financial report writing is still not simple, graphical accounting systems make the task easier and more approachable.

Graphical accounting improves decision support by providing a richer context for the data — by letting you combine textual data and charts, for example, as shown in Figure 3.2.

Often the textual and chart data can be linked to facilitate "what-if" analysis by letting you change either the textual data or the chart segments and see how the changes impact the view of the data you're working with. You can view documents and images attached to transactions, such as invoices, or to other records, such as employee cards, by displaying the linked image in a separate pop-up window. Graphical accounting also facilitates drilldown by letting you navigate from summary balance windows to successive detail transaction windows to trace data back to original source entries or documents.

Figure 3.2: A Graphical Decision Support Interface

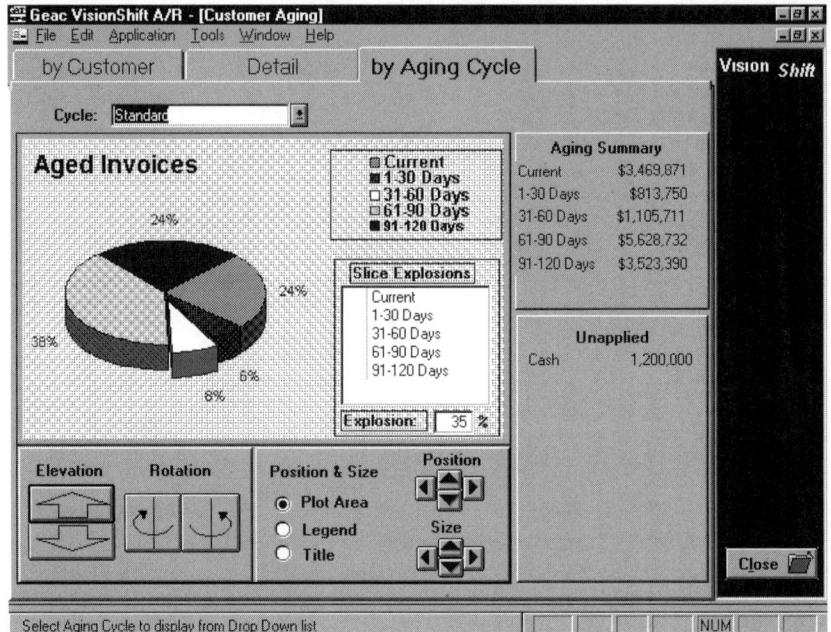

© Geac VisionShift 1996

ScreenHelp: This screen helps the user visualize the aging analysis in the VisionShift accounts receivable package from Geac VisionShift. The tab folder allows customer aging balances to be analyzed and charted by customer, by invoice detail, or by a user-selected aging cycle. The use of a colorful pie chart makes the aging breakdown easy to assimilate, particularly when it is combined with supplementary tables that summarize the debt and unapplied cash balances.

Graphical accounting also benefits from ease-of-use features that are simply a part of the underlying GUI functionality. These include

- ever-present menu bars with pull-down multilevel menu systems that avoid the up, down, and roundabout navigation of old-style hierarchical or ring menu systems

- multiple document interfaces (MDI), which let you display more than one form at time so you can easily compare and contrast information

- context-sensitive toolbars that gather the most frequently used functions onto a button bar so they are never more than a mouse click away

- drag-and-drop functionality for visually building relationships between data, such as linking an address record to an account record, using only the mouse

Graphical accounting does have some drawbacks. First, you need to have the supporting GUI installed on your computers. For most desktop users this is not a problem, but it does mean that users of terminals need to be converted to PCs or workstations of some sort. Even graphical accounting based on Microsoft Windows is not completely homogeneous, so for one reason or another, some vendors have deviated from the standards here and there, with uneven results. Look for vendor certifications, such as the Microsoft "Designed For Windows 95" logo, to indicate a minimum level of common standards.

Graphical accounting is not proven to be better than traditional nongraphical systems for rapid, heads-down transaction entry. Proficient users of nongraphical systems may need time to become familiar with how to use the GUI or a mouse — the double-click often takes a while to get the hang of. Some accounting systems marketed as graphical are nothing but emperors with no clothes — so-called "screen scrapes" that are ports of old mainframe-like screens or simply wild deviations from accepted GUI standards. You should avoid these. Some graphical accounting systems provide better keyboard support than others through the use of accelerator keys — function keys or key combinations that give users alternative ways of accessing functions. As users become proficient, most find using accelerator keys faster than using the mouse.

On the whole, graphical accounting is a major step for accounting applications in terms of usability and consistency. It is responsible for introducing many functional innovations for navigating and maintaining complex accounting structures, drilling through data, presenting data using more intuitive tabular or chart formats, and constructing and presenting reports. As you evaluate a new accounting system, you should view the GUI with the same critical eye you use on other aspects of the system. After all, the GUI is the part of the system design the accounting operators will face and use every day.

Is Graphical Accounting Just About Using a Colorful Interface?

It isn't just the GUI widgets or the intelligent use of color and layout that make graphical accounting software more attractive to users and easier to learn; it's also the increasing use of innovations such as cue cards, wizards, process maps, sophisticated help systems, and context-sensitive actions.

Cue cards are pop-up screens that provide tips and advice on how to use the screen you are in or what to do next once you've completed processing in the current screen. Cue cards are an adjunct to the standard help system, which requires users to find the help they need by navigating an online manual. Cue cards are a useful and immediate form of help to remind users of the purpose of the screen they are working in.

Wizards take cue cards a step further by actually automating process steps using a series of pop-up guided prompts. Wizards walk the user through a particular process, such as building a report or managing a month-end process, by asking a series of simple questions and using the answers as parameters to specific program functions. Wizards are particularly helpful in automating a complex task to ensure it is managed in the right order and fully completed. Increasingly, wizards are being used to assist in the time-consuming installation and setup of accounting software as a means of reducing the cost of getting up and running.

Process maps show the steps in an accounting process visually, sparing the user from searching various menus for the functions to carry out the steps. In addition to providing a useful road map for common processes such as month-end closing or applying cash, process maps can act as another means for navigating and using system functions. To launch the relevant application function, the user can usually click the picture or box that represents the process step on the map. Process maps can also restrict the user's progress to a specific flow or track it by showing the user an on-screen checklist of the completed steps in the process.

Help systems are also becoming more sophisticated in graphical accounting packages. Many, such as those supported by Microsoft Windows 95, provide an online manual users can navigate in a number of ways, including using hierarchical content browsers and searching by keywords or other text combinations. In many cases, the online manual is more comprehensive than the printed manual, especially if users can annotate the online manual by adding their own site-specific help to the standard help text. Some accounting systems also provide pop-up context-sensitive help at the data field (entry box) level of graphical accounting forms; this help may also point users to other linked topics that could be of assistance.

Context-sensitive actions, usually invoked by clicking the right mouse button, are also becoming an important new facility in graphical accounting systems. Generally, clicking the right mouse button pops up a menu of options sensitive to the data currently displayed or selected. For example, if the user selects and clicks a specific row on an inquiry screen showing a list of vendor invoices, a pop-up menu may let the user view an image of the invoice. Alternatively, the user may be able to a select a range of data, such as a list of account balances, and then select a charting option from a pop-up menu to view the data as a graph. Instead of being static, the data becomes a live object with a range of useful manipulation options and viewing perspectives.

Wizards, cue cards, process maps, online help, and context-sensitive actions all improve the usability of an accounting package and make it easier to learn, which in turn helps reduce implementation time and cost as well as day-to-day processing errors.

What Are Some Visual Widgets Used in Graphical Accounting Systems?

Vendors build graphical application interfaces using a set of graphical controls or widgets, of which there are typically dozens available to developers in any graphical user interface. Widgets are becoming ever more sophisticated, and in some cases are really miniapplications or application components embedded in graphical forms to supply specific functionality — for example, to display a tabular grid of data or to show data in the form of a chart.

Figure 3.3 illustrates and describes some common graphical widgets available under Microsoft's Windows 95 GUI that application developers can use to display graphical data entry and query forms.

Figure 3.3: Some Microsoft Windows 95 Widgets

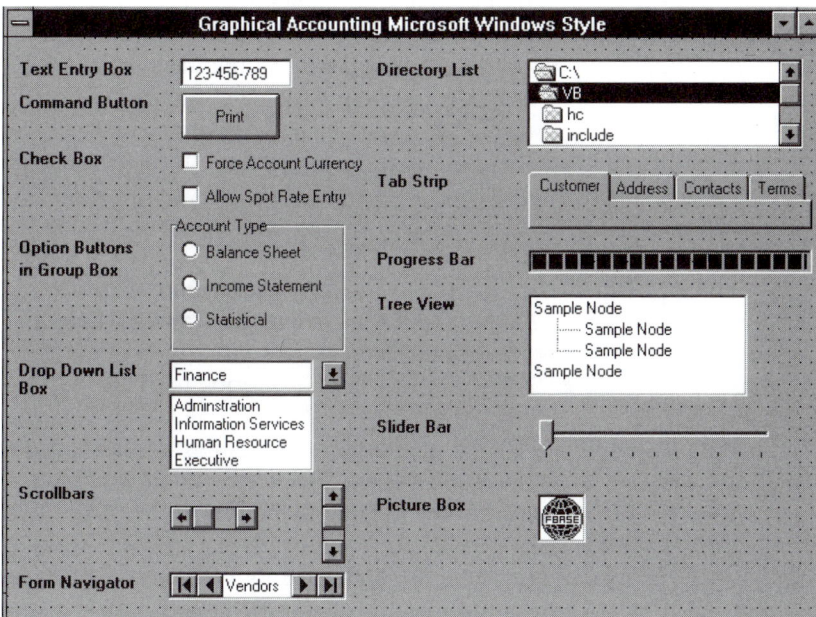

Widget	Description
Text entry box	Used to enter text such as account codes, dates, and descriptions. Usually you use the Tab key to move from box to box.
Command button	Clicking with the left mouse button launches a process, such as printing a report or displaying a new data entry or query form.
Check box	Clicking in a check box toggles the condition as applicable or not applicable.
Option buttons	Clicking a button sets the condition to one of the choices displayed.
Drop-down list box	Clicking the arrow displays a drop-down scrollable list of available choices. Clicking a choice selects it. Often you can key in the first few letters of a choice, and the list box will jump to the nearest match.
Scrollbars	Click the arrows to move data up or down or to the right or the left.
Form navigator	Click the VCR controls to move forward and backward or to the first or last record in a set.
Directory list	Displays a hierarchy, using icons to provide a visual differentiator between levels.
Tab strip	Click a tab to display a data entry or query form with content specific to the tab title.
Tree view	Displays a hierarchy. Double-clicking on a row at a given level expands or collapses the tree.
Progress bar	Shows the progress of a process such as posting a transaction or printing a report.
Slider bar	Click the slider to increase or decrease a value or range to be recorded.
Picture box	A picture may simply be ornamental, or it may indicate a hotspot. Clicking a hotspot initiates a process or displays another entry or query form.

Conclusion

Microsoft Windows 95 is the GUI of choice for accounting applications for the foreseeable future. As Microsoft pushes the limits of Windows 95, we're likely to see both more innovation and more standards in the ways GUIs are used to show and manipulate business information. However, as Internet browsers become more pervasive and vendors Web-enable their accounting suites, it's likely that the GUI through which many people access their accounting system will be based on a browser such as NetScape's Navigator or the Microsoft Internet Explorer. For others, particularly those participating in accounting-related workflows, the GUI world may consist of a universal electronic in-box used to manage e-mail and route workflow items. For a discussion of workflow accounting and universal in-boxes, see Chapter 9.

QUESTIONS FOR VENDORS

1. Does the application provide both graphical and nongraphical versions of the client applications for faster data entry?

2. Can the application be either mouse-driven or keyboard-driven?

3. Can users press Tab or Enter to move between fields in Microsoft Windows screens or forms, and can they use the numeric keypad for faster data entry?

4. Is the application certified "Designed for Windows 95" by Microsoft?

5. Does the application make extensive use of cue cards, process maps, or process wizards for ease of use?

6. Does the application ship with a fully integrated online help system, preferably with context-sensitive pop-up help screens?

7. Does the application use "tree" browsers for navigating accounting structures and tab folders for visually organizing logically linked data?

8. Can the application be completely installed using a graphical set-up wizard?

9. Can the application present information in a variety of formats, such as tabular grids and charts?

10. Does the application support visual drilldowns from summary to detail data to navigate linked datasets?

Chapter 4

Spreadsheet Accounting

Spreadsheet Accounting 101

Spreadsheet accounting is the integration of spreadsheet software with accounting systems. Spreadsheets are fast becoming a universal front- and back-end to all accounting systems. By offering direct access to accounting data, spreadsheets make error-prone, unaudited rekeying of data a thing of the past. The latest client/server accounting packages include direct transfer of data to and from spreadsheets without the need for messy file exports and imports. Spreadsheets themselves include new features such as data drilldown, report outlines, pivot tables, workbooks, and e-mail connectivity that make them even more suitable for financial reporting and analysis. And if this isn't enough, specialist vendors have stepped in to make spreadsheets even more attractive as a valuable accounting system add-in.

Spreadsheet software is one of the most useful tools on the accounting desktop — it could even be argued that the spreadsheet is too useful. The comfort of the row, column, and formula paradigm has meant that spreadsheets are often used for jobs for which other tools, such as databases or word processors, are better suited. Spreadsheets are almost certainly the world's most popular financial reporting tool, even though they were not originally designed to do that job. Consequently, many accounting systems are covered by a patchwork of spreadsheet "band-aids" used to manage consolidations, currency translations, financial statement production, and so forth.

Spreadsheets have come to be used this way because, in the past, accounting systems generally didn't offer easy-to-use reporting tools. The easiest way to get the information you wanted and format it in a way management would accept was to put it into a spreadsheet and "massage" it into shape. This process involved messy file transfers out of the accounting system and into the spreadsheet, or worse, wholesale rekeying of data into a multitude of linked worksheets. It wasn't uncommon for large corporations to devote a small army of people to this task. Once in the spreadsheet, the data had to be formatted for the spreadsheet rows and columns using macros written in the spreadsheet's programming language. As a result, closing fiscal periods could take weeks rather than days or hours. The process was cumbersome, time-consuming, error-prone, and largely unaudited, so it's no wonder that figures in financial reports were often out of step with the actual numbers in the accounting system.

Today, spreadsheet accounting has changed significantly, thanks to improvements in the integration of spreadsheets with accounting data, enhanced sophistication in the spreadsheets themselves, and the release of specialized spreadsheet accounting tools.

Spreadsheet Integration

Some years ago, the leading spreadsheet packages, Lotus 1-2-3 and Microsoft Excel, took a great leap forward: they added tools that let the spreadsheet directly access data stored in databases. In Lotus 1-2-3, the function was called DataLens, and in Excel it was an add-on called Q+E. From that point on, assuming your accounting data was stored in a file manager such as Btrieve or an SQL database, you could pull data directly from the accounting system into your spreadsheet without rekeying or parsing. With these products, you could establish a connection to a database from within a spreadsheet, query the database using SQL or a similar query language, and load the data directly into specified spreadsheet cells.

This was a major productivity improvement, and today no popular spreadsheet product is without integrated database access. Instead of requiring two steps — data export from the accounting system and file import into the spreadsheet — direct database access lets you define from within the spreadsheet simple queries that extract data from the accounting database and put it directly into spreadsheet cells. Once you define a query, you can save it with the spreadsheet to create a hot link to the database. You can then rerun the query as needed to update the spreadsheet report. Although this data access capability does let you define relatively complex queries, usually through SQL, it is fair to say that its use is limited to extracting lists of data rather than formatting complex, matrix-based financial statements.

More recently, spreadsheet integration has been extended, particularly through Microsoft's dynamic data exchange (DDE) and Object Linking and Embedding (OLE) technology. DDE is used to establish a channel of communication between two applications — one acting as a DDE client and the other as a DDE server— so that data can be passed between them. OLE lets you "embed" a spreadsheet into an accounting application so the spreadsheet software can load and manage the accounting application as a

EVALUATION HEADS-UP

- All modern graphical spreadsheets, such as Microsoft Excel, Lotus 1-2-3, and Corel Quattro Pro, can now access data in relational databases directly by using middleware such as Microsoft's ODBC. All client/server accounting data is therefore easily accessible from a spreadsheet tool and should not require rekeying or error-prone import and export.

- Many accounting systems can now pass data to and receive it from spreadsheets by using Microsoft's DDE or OLE technology. This capability makes it easier to initiate transactions in spreadsheets, such as entry of budget numbers or T&E expense timesheets, and then pass them directly to accounting modules for posting under the control of the accounting systems.

- Many accounting systems offer one-click toolbar-button data transfers into spreadsheets. These functions let you easily load report or query result data directly into worksheet templates for manipulation, analysis, or publication. OLE 2.0-compliant accounting applications also let you load and use spreadsheet software from an accounting screen and pass data to and from the spreadsheet.

- Because spreadsheet packages such as Microsoft Excel are e-mail-enabled, passing accounting data to a worksheet can be an easy way of distributing reports electronically if the accounting system itself is not e-mail-enabled.

- A number of accounting integration add-on products are available for the leading spreadsheet packages to add specialized data analysis and reporting functions for use directly against third-party accounting system data. Products such as F9 from Synex and MV Analyst from Timeline extend the functionality of the spreadsheet host and often include direct links to a number of popular accounting systems for easier data access and retrieval.

kind of subsession within the main accounting session. Application embedding makes it much easier for accounting applications to pass data into spreadsheet templates or receive data from them.

Through DDE or OLE, spreadsheets can also be used to manage the initiation of accounting transactions such as budgets or travel and entertainment (T&E) expenses. A spreadsheet is a natural format for manipulating budget data and collecting T&E trip data. You can enter data into a spreadsheet and then automatically transfer it to the accounting system using DDE or OLE. In some cases, a specially formatted spreadsheet is loaded directly with data from within the OLE-compliant accounting system software. When you save the spreadsheet data, it's passed to the accounting system's standard posting routine, which posts it to the accounting database subject to the same security and audit trail considerations as any other accounting transaction. This approach combines the benefits of spreadsheet data entry with automated, secure, and audited transaction posting under accounting system control.

Spreadsheet integration is not limited to relational databases. Many multidimensional databases are front-ended by spreadsheet-based applets that let users extract data directly into worksheets from the database engine. The worksheet/multidimensional database combination is particularly powerful because the source data is well structured to take advantage of worksheet innovations such as row/column pivoting and outliner-based drilldowns. These benefits add to the attraction of using multidimensional databases as the foundation for a financial reporting and decision support system.

Spreadsheet Reporting

Although spreadsheets may appear to be ideal for financial reporting, in practice it's very difficult to build a sophisticated, production-level, transaction-based financial report writer inside a spreadsheet. Nevertheless, some vendors, such as PeopleSoft, with its nVision report writer, have managed to do just that. More commonly, spreadsheets are used as a destination for financial reports that the accounting system's own internal reporting engine has already produced. Or they are used to feed data into the consolidation reporting process, which is why financial reporting vendors, such as FRx Software, allow spreadsheets as a data source for their financial reports. Spreadsheet add-ins such as the one shown in Figure 4.1 have made the process of building and managing financial reports from the desktop even easier.

ScreenHelp: This screen shows Timeline's MetaView Analyst spreadsheet add-in operating within a Microsoft Excel worksheet. The screen displays a MetaView Analyst wizard being used to gather data from a source accounting system for use by an income statement report template that is formatting the underlying Excel worksheet. The MetaView Analyst add-in includes built-in links for accessing data directly from many popular accounting systems.

One popular use for spreadsheets is to improve the presentation of financial reports for internal management or external analysts. Spreadsheets

Figure 4.1: Spreadsheet Add-ins Make Building
Financial Reports Easier

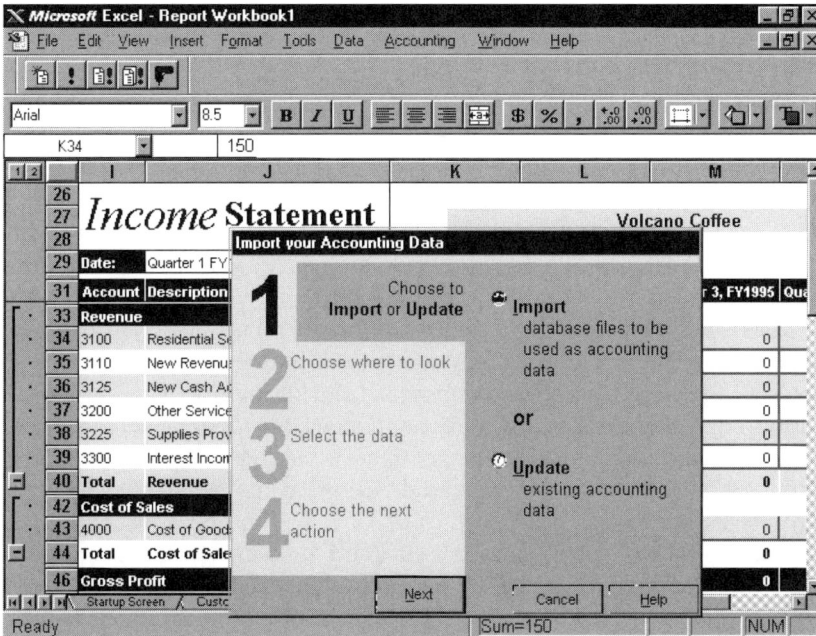

© Microsoft Corp. 1996, Timeline Inc. 1996

running under Microsoft Windows 3.1x or Windows 95 have rich data presentation features, including variable fonts, shading, colors, borders, headers
and footers, and embedded graphic images for incorporating corporate
logos. The one-click data transfer facility offered by many accounting systems can also be used to transfer finished financial statements directly into
spreadsheets. For example, by clicking a button on a toolbar in the accounting system, a user can load a spreadsheet package and automatically transfer data, correctly formatted, from an on-screen financial report into a
spreadsheet. From within the spreadsheet, the user can then save the fresh
report and print it or e-mail it to a distribution list of recipients.

Spreadsheets were designed for analyzing financial data, but new features such as report outlines and pivot tables have added significant new
analytical capabilities to assist with financial reporting. Report outlines
make it easy to expand and collapse the data represented in a spreadsheet
by intelligently dividing the report data into a hierarchy of levels. By clicking outline buttons to the side of the spreadsheet, you can collapse a report
to see a summary view or expand it to uncover more detail about each
report line. Pivot tables let you rotate data that includes multiple information

dimensions (such as company, account, period, and department) within a spreadsheet. With this feature, you can manipulate any valid combination of dimensions by changing the row or column dimension to reflect the view of the data you want to see. By extracting data directly from the ac-counting database or using one-click report transfers, you can then apply outlines or pivot tables from within the spreadsheet to view the report data in a myriad of new ways.

Spreadsheets such as Microsoft Excel are e-mail-enabled, so you can add report routing to accounting applications. After transferring data into an Excel spreadsheet, you can attach the spreadsheet to an e-mail message and route it to individual users or user groups included in the e-mail system directory. On Excel's File menu, this functionality is provided by a simple Send option.

Many companies like to prepare a month-end report pack for review by upper management. This pack may contain conventional reports, charts, and notes that present a complete snapshot of the business activity during the period. All reports, charts, and notes are managed in separate spreadsheets, which are then combined into a container or package called a workbook. The workbook package uses tabs to divide the contents so that clicking a tab reveals the underlying report, chart, or note. Workbook packages can also be routed via e-mail by using a standard Send menu function, providing a way to automate and electronically distribute the month-end report pack.

Conclusion

Spreadsheets are already a universal front- and back-end to accounting systems, and there is no reason to doubt that even tighter integration between spreadsheets and accounting systems lies ahead. We can expect to see

- native spreadsheet files (preserving report formatting and formulas when imported) being a standard output choice for financial report writers in every level of an accounting system

- native spreadsheet files being used to create packages of report data for e-mailing to remote or mobile users for local analysis and decision support

- more use of spreadsheet objects within accounting applications for entering data such as timesheets and budgets

- development of specific metadata layers in spreadsheets to allow faster and more intelligent connections to accounting databases

- increased use of spreadsheets as data collection and consolidation conduits — for example, for collecting summary business unit information, consolidating the data at regional or headquarters sites, and feeding the data into consolidation systems or general ledgers

QUESTIONS FOR VENDORS

1. Does the application support a one-click toolbar button transfer from the accounting system to a popular spreadsheet such as Microsoft Excel or Lotus 1-2-3?

2. How sophisticated is the transfer function — can it transfer only text and numbers, or does it include, for example, formatting tags and formulas?

3. Can the accounting data be accessed directly from a spreadsheet and still be subjected to the same security constraints as a regular accounting user?

4. Can the application support the automatic transfer of data from a spreadsheet to the accounting database without user intervention?

5. Can a spreadsheet be launched from within an accounting application session using OLE technology?

6. Does the package support popular worksheet formats such as Excel.XLS and Lotus.WKS as default output formats for accounting reports and queries?

Chapter 5

Adaptable Accounting

Adaptable Accounting 101

Adaptable accounting refers to accounting software that can be easily adapted or customized to fit your business processes. No generic, packaged software application can hope to offer a perfect fit to any particular enterprise's business needs, especially because those needs change — frequently in an unanticipated direction — over the system's lifecycle. Many businesses try to hedge that risk by selecting software that offers some degree of adaptability.

Adaptability can take many forms and may exist at varying levels of application functionality. Accounting software is often asked to adapt to

- the different computing platforms used throughout the organization

- the specific terminology and data capture needs of local business units

- a variety of business rules put in place by governmental and regulatory entities

- differences in the implementation of specific business processes

- the need to integrate accounting with other business systems

- the demand for add-on functionality not provided by the application vendor

Such adaptability is in addition to the flexibility already built in to the accounting application in terms of its user-defined account code block (e.g., the account code block segmenting), transaction analysis, and reporting structures (e.g., the financial report writer's row and column formatting capabilities).

Platform Adaptability

Often termed portability, platform adaptability is the application's ability to function on different combinations of computing platforms. From a client/ server perspective, the platforms are those supported by the various application layers and services. Some of the platform choices for the application layers are listed below.

Layer	Popular Choices
User interface	Microsoft Windows, MacOS, OSG/Motif, Character
Network	TCP/IP, Novell IPX/SPX, Microsoft NetBEUI
Database	Oracle, Sybase, Informix, IBM DB2, Microsoft SQL Server
Server OS	Unix, IBM OS/400, Microsoft NT, Novell NetWare

Accounting applications that offer a range of platform options may let you mix and match those options. Such flexibility can be very important to enterprise-level businesses whose business units use many different combinations of platforms. On the other hand, applications that focus on a specific platform are more likely to take better advantage of it. For example, a number of middle-market accounting vendors use the applications of Microsoft's BackOffice platform as the foundation for their accounting systems. These products are more likely to get the most value from this platform's rapidly changing and improving function set.

Platform adaptability is often a by-product of the programming language used to write the accounting application. If the application is written in a cross-platform fourth-generation language (4GL), it inherits the ability to operate across that same range of platforms. Consequently, it's useful to know what technology was used to develop the accounting package, because that technology may enhance or limit the application's adaptability to other platforms. The vendor profiles in Part 2 of this book cover the technology and platform choices for 54 vendors.

Look-and-Feel Adaptability

Many businesses need only enough adaptability to be able to customize the look and feel of the accounting application. Look-and-feel adaptability means being able to

- change the terminology used on screen data-entry and query forms

- hide functionality or specific data from view on specific screen forms

EVALUATION HEADS-UP

- These days, you shouldn't need an application that includes a source code license, unless you really want full control over every aspect of the system and the burden of maintaining all your changes yourself. Many vendors no longer even offer source code licenses. Instead, consider applications written in popular 4GLs, such as Microsoft's Access or Visual Basic, Centura SQL Windows, Compuware Uniface, Powersoft PowerBuilder, or Smalltalk, that are accessible, through the 4GL environment, for user customization.

- Many client/server accounting applications can be deployed on a broad combination of GUI, database, and server platforms. Decide early on whether you need multiplatform deployment support or whether you want to focus on one specific platform combination.

- Is the basic adaptability you need included in the package, or will you want to customize the package and add functionality of your own? If the former, look for packages with at least a user-accessible data dictionary and forms customization toolkit. If the latter, choose one written in a 4GL or one that includes an IDE.

- Does your IS department have expertise in a particular 4GL or programming language? If it does, look for an accounting package based on similar technology to leverage your staff's skills. Although a 4GL-based accounting package may not perform as well as a package written in native C or Cobol, it may be much easier to customize and have a more "state of the art" look, feel, and function set.

- If you need to integrate your accounting system with other packages, look for sophisticated data import, export, and mapping capabilities and support for application integration technology such as Microsoft OLE. For linking to other distribution and manufacturing systems, look for built-in support of EDI and EFT file formats or support for emerging application integration standards, such as those promoted by the OAG.

- To assemble your accounting application around the specifics of your own particular business processes, look for software architected around a workflow paradigm. Such software is much more process-centric and adaptable to changes in process flows.

- add new screen form data entry fields for use by certain users

- reduce the scope of the application by hiding menu options or buttons

- translate or annotate prompt, help, and error messages to suit local usage

A certain amount of look-and-feel customization is achievable through an application's security system, which associates certain options with particular users or groups of users. The security system can be used to

- reduce the number of menu options available to users to slim down the functionality or prevent access to sensitive functions

- restrict the use of toolbar or form buttons that let users enter data or edit or delete existing data

- hide or display certain fields on data entry or query forms

Applications that have a user-accessible data dictionary and a screen-form customization toolkit make it easy to further customize the look and feel. A data dictionary stores a range of information, known as metadata, about the application's forms and fields. Metadata includes information ranging from the type and length of a data entry field to the field's color. For example, an amount field may be defined as a numeric-only entry field that is 10 digits long, formatted to two decimal places, and displayed in red if the number it contains is negative. The application uses this information to generate the screen forms users see and work with. With a form customization toolkit, system managers can call up templates for standard application forms and edit them by changing the template attributes in the data dictionary.

Data dictionaries are a powerful way for system managers to modify aspects of the application's internal data definitions, such as

- the text displayed on forms as a label for a particular data entry field

- the type, length, and validation attributes of a data entry field

- any special rules associated with the use of a data entry field

- the prompt, error, and help messages associated with a data entry field

- the color of fields on a form and whether they are visible to the user

Another benefit of data dictionaries is that they make it very easy to translate software into foreign languages — you need only provide an alternative data dictionary with all text, such as screen field labels and error messages, in the new language. Because the dictionary is the central repository of application metadata, once edited, a data item will appear and function in the same way wherever it's used in the application.

A form customization toolkit provides a repository of form templates that can be edited to suit your look-and-feel requirements. Figure 5.1 shows one vendor's form customization toolkit. This form-based approach lets you manage everything about a form and its data using an on-screen graphical editing tool. Changes are usually limited to a specific form and may be visible only to certain users, to allow different users to see different versions of the same form, driven by the security system. Although more visual than using a data dictionary, the form-based approach may be less flexible because changes may not always ripple through to other forms that reuse the same data.

Figure 5.1: Solomon IV for Windows' Form Customization Toolkit

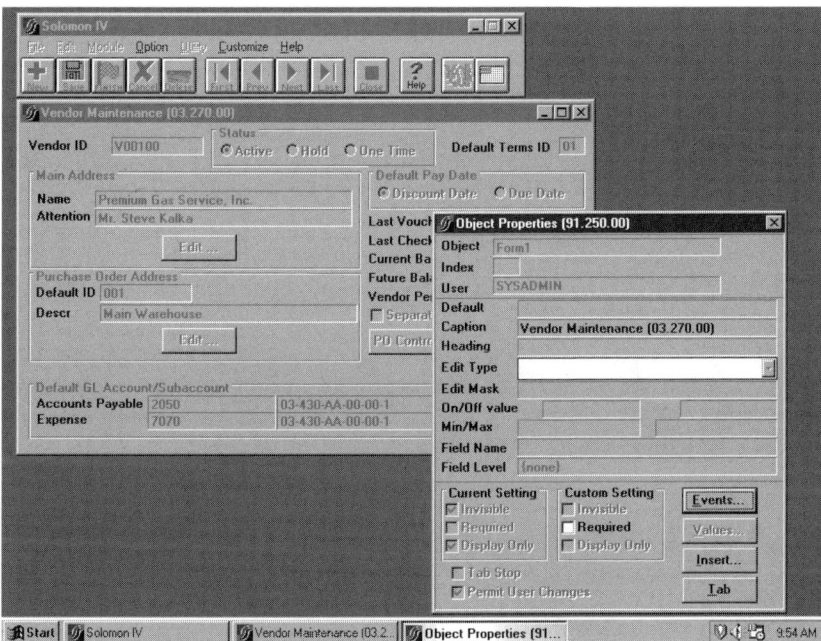

© Solomon Software 1996

ScreenHelp: This screen shows a vendor maintenance form from Solomon Software's Solomon IV For Windows. Users can customize a form and all the controls on it through the Object Properties screen, which displays user-adaptable attributes stored in the data dictionary. The Events button lets the user define Microsoft Visual Basic-like code to be attached to the form or the controls for delivering more sophisticated functional behavior.

Functional Adaptability

Accounting packages may support various levels of functional adaptability. For example, an application may let you

- customize existing application business rules and logic
- "hook-in" new business rules and logic to existing functions
- add on wholly new functions and processing logic

Applications designed with adaptability in mind often logically — and physically — separate their business rules and logic from the rest of the application code. This is particularly likely to be the case when the application is partitioned for three- or n-tier client/server deployment. The benefit of this separation is that it often makes it easier for system managers to access, change, or add value to the business rules and logic.

Many client/server applications keep business rules and logic in stored procedures or triggers, which are maintained as objects in the database. Unless they are in some way encrypted to prevent access, these objects can be inspected and edited by any proficient programmer or analyst. One drawback to putting the rules and logic into stored procedures or triggers is that when the vendor releases an application upgrade, any amended stored procedures and triggers must be reloaded into the database and tested at the installation site. This requirement adds time and complexity to the upgrade process.

In the past, many packages included so-called user hooks in their software where users could insert alternative or additional business-specific code. A typical place for a user hook would be where an accounting application calculates discount as part of an order line-entry program. Because discount policies vary widely across industries, the vendor-provided options for calculating discounts might not be suitable for use in a specific business. In this case, the user could insert a new piece of discount-calculation code to be used instead of the original code. Applications that provided user hooks for adaptability might scatter hundreds of them through the application.

In today's more object-oriented and event-driven accounting system designs, user hooks are replaced by user-definable event processing. Under this approach, the screen forms used for data entry and query — and the data fields on those forms — are associated with a rich set of events. Events are the pieces of code the system executes in response to a specific action, such as clicking the mouse, displaying or closing a form, or pressing the Tab key to move between fields on a form. Users can expose and modify the code for events to suit business needs. When software updates are released, users can have the software produce exception reports that help ensure that the modified event code is carried into the new version.

The ultimate functional adaptability is offered by accounting applications that let the user build and integrate completely new functions and the database tables to support them. This type of adaptability requires sophisticated application generation capabilities, including the ability to

- define new database tables to store the data for the new function

- define new screen forms to enter data into and query it from the new tables

- link the new forms into menus or other function-selection options

- provide messages and help for the new functions

- register the new tables and forms with the application so they can be documented and secured

For example, you might want to link a new sales contact module to your accōunts receivables module so sales staff can store information about the customer. The new module might include information about the sales rep, the customer's product preferences, and a record of customer contacts. Such a module might require that a handful of new database tables, data entry forms, and reports be defined and linked to the existing accounts receivable customer maintenance form.

These tasks are easy to do using an accounting product that includes some form of Integrated Development Environment (IDE). The IDE is usually delivered as an optional module that simply adds additional functions to the application's main menu. The IDE uses visual table- or form-building windows, rather than a programming or macro language, to let the user

- access the application's data dictionary to maintain existing tables and forms

- create new tables in the database and relate them to other existing tables

- create new forms for maintaining data in new and existing tables

- create new code that can be attached to existing or new forms and data entry fields

- link the forms to new menu items or toolbar buttons so they can be accessed by users

Once defined in the IDE, the new functionality is made available simply by closing and reopening the application. What's more, such changes can often be achieved by knowledgeable users themselves, as typically no actual coding is required and wizards help with much of the customization.

Integration Adaptability

Integration adaptability is the accounting software's ability to integrate with other business systems. Usually the integration adaptability depends largely on the software's data import and export functionality. Accounting applications typically must integrate with many other business systems, including

- budgeting, payroll, and fixed-asset systems

- reporting, consolidation, and data warehouse systems

- spreadsheet-based timesheet or T&E expense systems

- manufacturing and distribution systems

- e-mail, workflow, and document management systems

To make integration as easy as possible, the application should include import and export functions that provide sophisticated data mapping and conversion capabilities so that incoming data is mapped correctly into the accounting system's data tables and outgoing data is fed in formats the receiving systems can handle. Support for technology such as Microsoft dynamic data exchange (DDE) or Object Linking and Embedding (OLE) can help with application integration, especially with data transfer to and from spreadsheets and other desktop applications. Support for output of business documents such as purchase orders, invoices, and checks in electronic data interchange (EDI) or electronic funds transfer (EFT) formats help integrate with third-party distribution and payment systems. And support for newly emerging standards, such as those promoted by the Open Applications Group (OAG), helps integrate heterogeneous financial, distribution, and manufacturing systems through use of business document interface specifications. E-mail, workflow, and document management systems depend on support for other commercial technology standards such as

- Microsoft's Mail Applications Programming Interface (MAPI) and the Internet's Simple Mail Transfer Protocol (SMTP) electronic mail protocols

- standards emerging from the Workflow Management Coalition

- document scanning standards, such as TWAIN for desktop imaging

Process Adaptability

The handful of available accounting systems that are designed around a workflow architecture can also offer process adaptability, or the ability of the application to be assembled around a specific implementation of a business process. For example, the business process for acquiring a fixed asset or approving a vendor invoice differs from organization to organization. Process adaptability helps ensure that a business is not forced, by the design of the software, to manage this process differently from the way the business is accustomed to doing it. To provide process adaptability, accounting software must be designed around a workflow paradigm and offer exceptional functional granularity, the ability to define process flows, and the ability to manage, modify, and monitor the events, rules, and routings associated with those processes. Process adaptability is becoming a very important aspect of the functionality demanded of accounting systems as more and more businesses attempt business process reengineering.

Process adaptability lets you literally reassemble the accounting functionality around a particular process flow by "mapping" software functions onto the steps in a process. These workflows can then be represented to users graphically, as in Figure 5.2.

Figure 5.2: Graphical Representation of User-Specific Process Flows

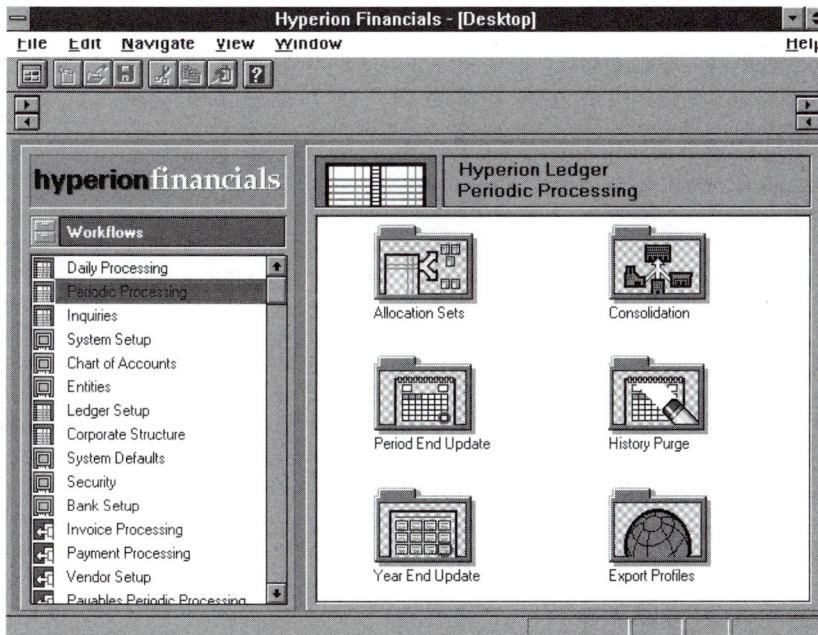

© Hyperion Software 1996

ScreenHelp: This screen shows a desktop form from Hyperion Software's Hyperion Ledger. On the left is a scrolling list of the workflows in which the current user can participate. Highlighting a workflow displays a graphical view of the process steps and lets the user click on a specific step to carry out the process task. This process adaptability helps to define the functional world in which users live when they use the system.

Because these process workflows can then be associated with specific users, different users can see completely different views of the functionality offered by an accounting module, and applications can be assembled around the way users actually work instead of the way the application programmer thinks they should work.

Object-Oriented Adaptability

The few accounting applications that are fully object-oriented also add new potential for adaptability. These systems are designed using frameworks of business objects. Business objects are self contained "chunks" of application functionality that encapsulate both data (attributes) and the tools (methods) for working with the data. Business objects generally map to real-world people, places, or things, such as customers, warehouses, or inventory items, and are represented as such to users by graphical icons on a desktop. Clicking the icon "launches" the object and makes its data and functionality visible and usable.

A framework is a set of common services and business objects that encourage users to extend the application's generic functionality by adding new, specialized functions. For example, the part of a framework that manages the storage, retrieval, and display of textual application messages may be extended to let the user record and play back spoken messages. Because the framework provides the underlying "plumbing" for all application messages, whether text or spoken, the new messaging functionality would be available to all application business objects that use the framework's messaging structures. This type of flexibility makes it easy to adapt an accounting application to integrate new technology as it becomes available.

There are two key advantages to object-oriented applications with regard to adaptability:

- You can create new business objects based on existing objects. A new object may inherit some or all of the attributes and methods of the object on which it's based.

- You can change the scope and behavior of business objects (such as a new object you've created) to suit your needs by adding new attributes and methods or editing the existing attributes and methods.

These features of object-oriented applications make it easy to, say, create a new type of cash customer object called CustomerWalkIn that requires no credit check or address. You can use the generic customer business object as a template for the new specialized customer object, which will inherit all its ancestor's characteristics. You can then switch off, modify, or add to these characteristics as required to make your new customer object behave the way you want it to.

Business objects and frameworks have a big advantage when it comes to managing the software upgrade process. Accounting software that is not built from business objects generally forces users to upgrade their whole system — or at least a whole module — at a time. This is the case even if only a small part of the system has been upgraded or the user needs only certain of the system enhancements or new features provided. However, a system built from business objects lets users upgrade object by object or slot new objects into the existing framework without needing to undertake a full system upgrade. Systems built from business objects will make it practical for users to manage system upgrades electronically by downloading new or enhanced business objects from software repositories maintained on the vendor's extranet. Upgrades may also be cheaper because businesses will be able to purchase feature upgrades à la carte by selecting only those the business can immediately put to use.

Users who customize business-object-based systems by building their own business objects using capabilities inherited from vendor-supplied ancestor objects should also have an easier upgrade process. When an ancestor object is upgraded on the user's system, the user's customized object will continue to work without change. The user may subsequently be able to run some form of re-inheritance routine to cause the upgraded ancestor object's new attributes and methods to ripple through to the user's customized objects. Thus, the customized object can continue to run as is, or it can be enhanced by allowing it to inherit new capabilities from the upgraded ancestor object.

Source Code vs. 4GL vs. IDE

In the past, many businesses wanted the source code for their accounting applications so they could have full control over the customization of the application, continue using the software even if the vendor went under, or raise the level of their relationship with the vendor. But source code licenses could be very expensive, and modifying source code could be tricky. In addition, changing the code in-house made upgrading the software difficult, if not impossible, and essentially shifted the responsibility for tracking and maintaining the modified software from the vendor to the user. Today, due to improvements in the adaptability of accounting software, few businesses need or want the cost and responsibility of making and maintaining changes to the source code.

A number of vendors have largely eliminated the customer's need for source code by using 4GLs or by providing built-in IDEs. Using a 4GL benefits software vendors by letting them

- speed initial development cycles and deliver revisions more rapidly

- deliver features and functions built in to the 4GL that would be difficult to justify if they had to be coded from scratch

- leverage the native portability of the 4GL to cross-computing platforms (GUIs, servers, and databases)

Accounting users benefit from the vendor's use of a 4GL because

- there is look-and-feel compatibility with applications the user may have written in-house using the same 4GL product

- they can leverage 4GL skills they've developed internally by using them to customize the accounting application

- the interface and other technology delivered by the 4GL vendor enhance the value of the accounting application

- the 4GL lets the vendors provide faster turnaround in response to customer wish lists and customization projects

Compared with applications written in more conventional programming languages such as C or C++, those written in a 4GL are often associated with less than optimum performance and high processing overhead, although these problems are being steadily eroded as 4GL products mature and are optimized more closely to specific platforms. Furthermore, these drawbacks may be overshadowed by the fact that application code built using a 4GL is accessible for customization by in-house programmers, who may be proficient in the language as a result of having used it to build other applications for the organization.

IDEs, on the other hand, are fully integrated into the accounting application and are not usually based on commercial 4GLs, but are specific to the package. Often, the entire accounting system is built using the same development environment. Although these environments are proprietary, they can be much more accounting- and application-aware than a generic, commercial 4GL can ever hope to be. Recently, IDEs such as the one illustrated in Figure 5.3 have come more and more to resemble other familiar and popular 4GL-type tools, such as Microsoft's Visual Basic or Borland's Delphi, by adopting similar feature sets and a similar look and feel. This similarity makes the IDE much more approachable to programmers familiar with the commercial tool on which the IDE is modeled.

Figure 5.3: Sample of an Integrated IDE

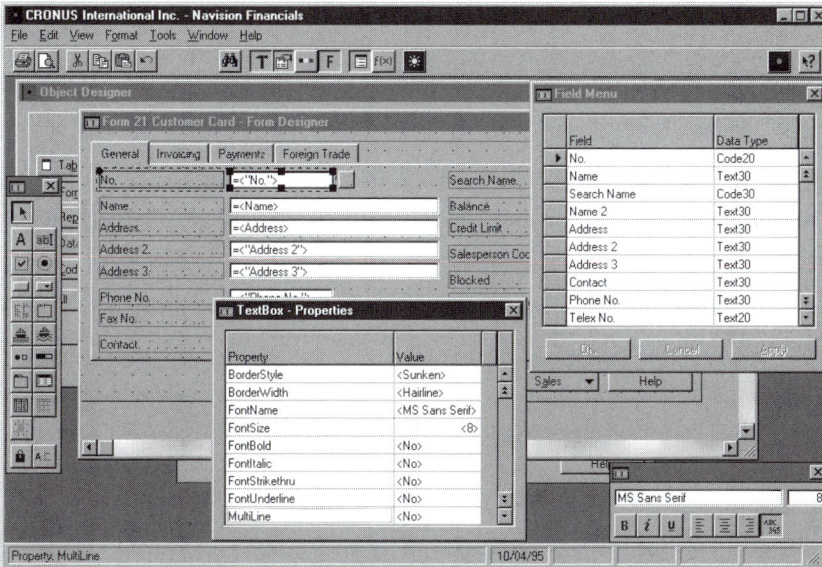

© Navision Software US Inc. 1996

ScreenHelp: This screen shows the IDE from Navision Software's Navision Financials in action. The IDE lets system managers define new accounting tables, build forms around those tables, and link the forms to the standard Navision Financials application using a set of customization tools familiar to anyone who knows 4GLs such as Microsoft Visual Basic or Microsoft Access.

Vendors who use an IDE to build their accounting applications may provide the IDE to their customers as part of the standard application deliverables. Using an IDE can be a revelation, especially to people used to source-code-level adaptability, and even to people familiar with 4GLs. IDEs allow systems to be changed quickly, easily, and on the fly, and they make the changes made or the new functionality created available almost immediately without requiring programming or recompilation. IDEs can be used to literally rebuild an accounting system from scratch or to surround an existing core of generic modules with a whole new set of business-specific modules. For multisite implementations of accounting systems across an enterprise, an IDE can be a big help to the IS department charged with customizing the package to the needs of different business units.

Conclusion

More and more vendors are building their systems using third-party 4GLs to provide a high level of adaptability, and this trend will continue as 4GLs become more sophisticated. Now that Microsoft licenses its Visual Basic for Applications (VBA) language, we can expect more vendors to use VBA as their built-in package customization tool rather than developing and maintaining their own proprietary languages.

As more packages move toward an object-based, component architecture, we should see adaptability moving down to the business-object level. As a consequence, users will be able to modify the behavior of individual functional objects and how they cooperate with other objects. Object-level adaptability will in turn make it easier to adapt accounting applications because changes can be isolated to the business objects that need them rather than having to be applied at a source-code level. Process adaptability, the next frontier in application customization, demands recognition of sophisticated workflow concepts in the package design.

QUESTIONS FOR VENDORS

1. Does the accounting application offer a customization tool, and is it based on a proprietary langauge or a third-party 4GL?

2. If based on a third-party 4GL, which version is supported?

3. If based on a proprietary language, is that language and its visual interface similar to other commercially popular 4GLs so the language will be easier to learn and use?

4. How extensive are the customization capabilities? Are they limited to the look and feel, or can you define new tables and forms and easily link them to the standard application menus?

5. What happens when the vendor releases a software upgrade, and how do upgrades affect modifications you've made?

6. When modifications have been made, can the application be used immediately, or does it need to be recompiled for the modifications to take effect?

7. Can you buy a library of third-party add-on modules built using the application's customization features or 4GL?

8. Can modifications be applied so that they are available only to selected users or user groups?

9. Is there a facility to automatically document the modifications made, check the integrity of modifications, and report design errors or exceptions?

Chapter 6

E-Mail Accounting

E-Mail Accounting 101

E-mail accounting is the integration of electronic mail capabilities with accounting software. Adding e-mail capability is one of the simplest and most productive enhancements you can make to an accounting system. E-mail enhancement lets the accounting system communicate with an e-mail system through a standard messaging protocol such as Microsoft's Mail Application Programming Interface (MAPI) or the Internet's Simple Mail Transfer Protocol (SMTP). An e-mail-enabled accounting application lets users

- compose messages to be sent to other users by e-mail from within the accounting application

- access the e-mail user directory for routing messages to individuals or groups

- attach other data, such as spreadsheets, for transmission with a message

Messages sent from within an accounting system appear in the receiver's e-mail in-box just like any other message, with or without attachments. Accounting systems can use electronic mail for such tasks as

- sending electronic notifications

- financial report distribution

- routing documents for approval

- packaging information as part of process workflows

Electronic Notification

Electronic notification is the use of triggers in the accounting database to automatically generate and disseminate information or alert messages by e-mail. Electronic notification merely uses e-mail as a routing mechanism for messages generated by notification functions built in to the accounting database. These notification systems can be based on simple thresholds attached to a value in a table, or they can use complex report-mining algorithms and workflow rules.

Electronic notification can send e-mail to inform specific accounting users that a particular event has taken place, such as a period close that requires all new entries to be booked to the new period. Managers can be alerted to exception conditions in the data, such as a project or departmental expense budget that has been exceeded or an inventory item that's reached the reorder point. Preprogrammed, rule-based report-mining software agents can interrogate an aging analysis report, determine that a customer has a large portion of debt more than 90 days old, and push a workflow item on the account sales representative to take corrective action. These are just some of the ways system designers can use electronic notification to add value to an accounting application.

Report Distribution

Printing, copying, and distributing paper-based reports become largely redundant when an accounting system is e-mail-enabled. Instead of making users print a hard copy of a query result or report, e-mail-enabled systems add another output option, the Send or Mail button (Figure 6.1). When a user presses the e-mail button, the report is produced the normal way but is stored as a file, which can be easily attached to an e-mail message. The user then composes a message, attaches the report file, and distributes it to an individual or group. Report distribution can be used to distribute packs of reports, such as a month-end pack, and the entire distribution process can be automated through the use of predefined report scheduling and routing rules.

ScreenHelp: This screen shows the ad-hoc query builder from Geac VisionShift's accounts payable module. The buttons at the foot of the form show the output options; note the Mail button, which sends the results of the query to the designated recipient(s) by e-mail.

Receiving users can review or print the report or upload it to a local database or spreadsheet for further analysis. This type of electronic report distribution extends the reach of the accounting data to nonaccounting managers and embraces mobile computing users by letting them download report data to their laptops and analyze the information locally.

Figure 6.1: An E-Mail-Enabled Query Builder

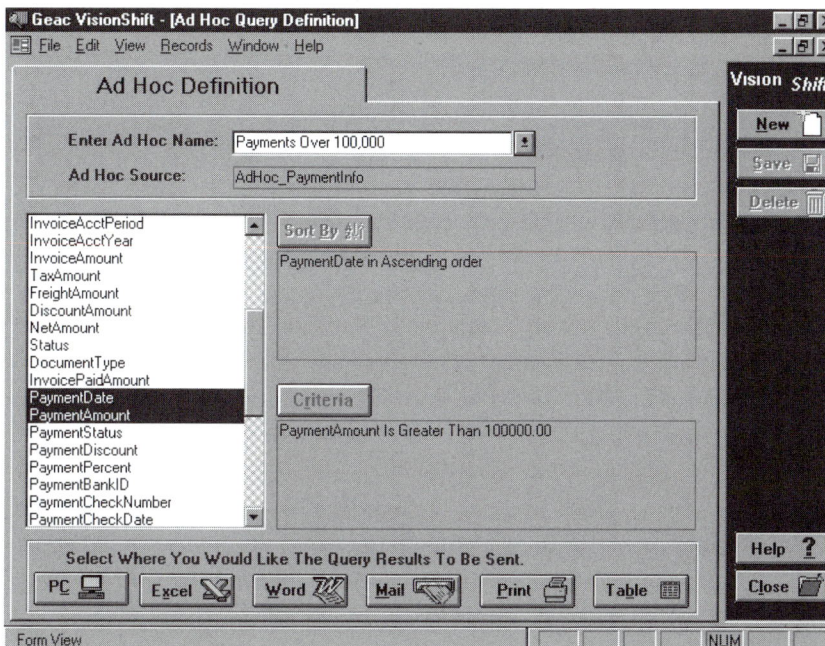

© Geac VisionShift 1996

Document Routing

E-mail can also be used to route documents as part of transaction workflows that may include annotating, editing, approving, or rejecting the document. Document-related accounting transactions such as sales orders, purchase requisitions and orders, project and T&E timesheets, and payment requests can be entered into the accounting system and then routed via e-mail for editing or approval. The documents can then be returned to the sender and the new amendments reflected in the document content or the approval or rejection decision communicated back to change the status of the document in the accounting system. With e-mail and document routing functionality, there should be few reasons for managing any accounting-related approval processes in a manual, paper-based mode.

Information Packaging

Information packaging lets users connected by e-mail participate in workflow processes. An information package is an attachment to an e-mail message that the recipient can manipulate to continue the workflow. The information package is a special kind of attachment that the receiving user

EVALUATION HEADS-UP

- E-mail-enabling an accounting system is one of the most productive "free" add-ons available from a client/server accounting system. Look for support for popular mail protocols such as Microsoft's MAPI or the Internet's SMTP.

- Check the level of e-mail enabling the vendor offers. Can you use e-mail to route reports for distribution? Can you use it to route transactions to automate approval processes? Does the application provide a notification or alert system that can be hooked into e-mail? Do the links to e-mail appear in logical places — for example, as output options in query and reporting functions or as a means for routing transaction approvals?

- Look for an accounting application that lets you use e-mail to send information packages containing data recipients can interact with (such as documents the recipient can approve or reject) or applets the recipient can use to manipulate data locally.

- For simple messaging between accounting-application users, e-mail integration may not be necessary. A simple database-based mail system may be all you need. At this level of messaging, a messaging infrastructure is not necessary.

or system can manipulate. Often the package contains both data and a pointer to a program executable, which may be stored locally or remotely. When the user double-clicks the package attachment, both the program and the data are loaded for action. The software required to run the program may not need to be loaded on the approver's PC, simplifying application maintenance and possibly reducing overall license costs.

One example of an information package is an order placed through a storefront on the Internet. The order information is packaged as a compressed and encrypted e-mail message attachment that the Internet storefront automatically sends to the vendor's Web server. There, software designed specifically to look for such packages opens the data packet, parses the data, and passes it to the accounting application's sales order-entry module as a bona fide, unapproved sales order. In this case, the use of an information package and an e-mail routing system effectively reengineers the sales order process.

Another example of an information package is a Java workflow applet routed by mail to let users participate in a transaction workflow. The applet is in the form of a business object that contains both data and functions for working with the data. The applet could be "pulled" down from an intranet server or sent as an attachment to an e-mail message. Upon receiving the message, a user "launches" the applet to, for example, view an invoice, add comments and annotations, and approve or reject the document. The applet is routed to multiple recipients until eventually the workflow reaches its end point and the transaction edits and additions, approvals or rejections are saved to the database.

In another case, a report "snapshot" database may be sent to remote users for local analysis. The report snapshot is not just a static view that recipients can scroll up and down, but a real minidatabase that includes all the transactions contributing to the final report. Consequently, recipients can drill down on report lines to uncover more detail, perhaps to the source transaction level. Recipients can also append the data to their own local database to build a true inception-to-date view of the information — for example, of sales performance to date or expenses accrued to date.

Information packaging is poised on the brink of a big future. It will become part of the way accounting systems manage business process work-flows as accounting applications are broken down into applets and the sophistication of messaging systems and electronic in-box technology increases over the next few years.

Conclusion

E-mail will certainly be integrated with all accounting systems within the next few years. The most dramatic impact will be on financial reporting, which is fast becoming a wholly paperless process. We can expect to see accounting systems communicating messages and information packages, using universal in-box packages such as Microsoft's Exchange or Novell's GroupWise product. These messages and packages will let nonaccounting users participate in accounting workflows, such as approving invoices, without having any actual accounting software on their desktop. As software-agent-driven report mining and notification systems become more common, e-mail will provide the means for proactively involving all kinds of enterprise users, at all levels, by routing the results of exception and opportunity information resulting from changes in the accounting database.

E-mail integration should also improve the interaction between vendors and users by providing a means of automatically notifying vendors of the occurrence of certain types of application errors or electronically distributing self-installing application updates.

QUESTIONS FOR VENDORS

1. Is the application genuinely e-mail-enabled in that it can communicate with external e-mail systems? Or does it only provide a simple means of sending messages to users who are logged on to the accounting system?

2. Is the accounting application itself really e-mail-enabled, or must it export accounting data to other e-mail-enabled applications, such as Microsoft Office applications, to take advantage of e-mail routing?

3. Which e-mail protocols are supported? Do they include protocols for transmitting the data across the Internet?

4. Which modules and functions can benefit from e-mail? Can financial reports and query results be distributed by e-mail? Can approval processes take advantage of e-mail routing?

5. Does the application generate notifications or alerts that can be delivered via e-mail?

Chapter 7

Image
Accounting

Image Accounting 101

Image accounting links image storage and retrieval technology with accounting software. An image is any graphical object (e.g., a photograph, chart, map, or diagram) that has been stored electronically, scanned into a database, or received in an electronic format, such as by fax. In this chapter, I use the term document to refer to formatted information that can be used as the basis for an accounting transaction, such as an invoice, check, goods receipt note, or purchase order. Whether the image is a picture or a transaction document, it is usually scanned from a paper original and then stored in an image format in a database so it can be viewed or manipulated on screen.

Image and document access in an accounting system adds value to the accounting information. Seeing a photograph of an employee in a human resource system or an assembly diagram of a manufactured item in a bill of materials system makes the viewed information easier to assimilate. Linking an original document such as an invoice or check to an accounting transaction provides the ultimate audit trail because the document is often what initiates the transaction. A number of accounting-system modules can benefit from the availability of images to supplement the standard text and numeric information. Here is a sampling.

Application Module	Type of Image
General ledger	Cash transfer and deposit slips, T&E expense timesheets
Accounts receivable	Sales invoices, customer statements, checks received
Accounts payable	Vendor invoices, checks paid, vendor terms and conditions
Purchasing	Purchase orders, vendor price lists, discount agreements
Inventory	Pictures of items, bill-of-material assemblies, warranties
Fixed assets	Pictures of assets, copies of insurance or title contracts
Human resources	Pictures of employees/applicants, resum,s, and professional/educational certificates

Adding imaging capabilities to your accounting system requires additional hardware and software, which may include

- scanning stations with scanning software and linked desktop or high-speed batch scanners

- indexing stations for tagging images with accounting-related data such as transaction numbers or account codes

- image manipulation stations for viewing and manipulating images outside of facilities provided in the accounting system

- image servers that manage the databases in which the images are stored and indexed

Input and Output Documents

When you think about documents in an accounting system context, you can identify two types — input documents and output documents — as illustrated in Figure 7.1. An input document is one that initiates an accounting transaction, such as a purchase order or a sales invoice. An output document is one generated by the accounting system, such as a vendor check, a customer statement, or a report (e.g., an audit trail listing or financial statement). Some output documents — checks, for example — become input documents in another accounting system.

Document management systems aim to reduce or eliminate the need for anything other than electronic or digital documents — in other words, to banish paper from accounting systems. In pursuit of this aim, documents are received electronically; are converted into electronic format as early in the transaction lifecycle as possible; or are never printed, but are simply created as electronic versions and stored in digital report libraries on dedicated report servers.

Figure 7.1: Input and Output Documents

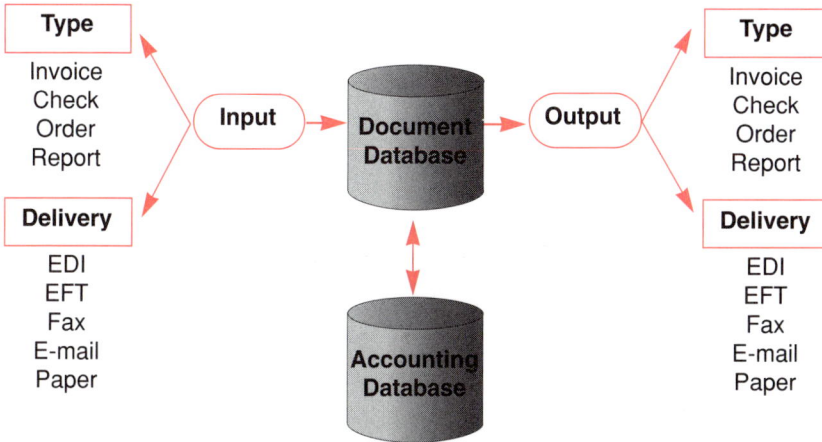

Despite the digital revolution going on around us, most accounting trans-actions are still rekeyed from paper documents mailed or faxed between businesses. Integrating these documents with the accounting transactions created from them involves the three-step process illustrated in Figure 7.2.

Figure 7.2: Three Steps to Digital Documents

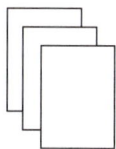

Step 1:
Receive
and batch
documents
for scanning

Step 2:
Scan
documents
into database

Step 3:
Tag or index
documents to
accounting
transactions

Step 1: Paper documents are received by mail or paper fax and batched together, ready for scanning.

Step 2: Batches of documents are scanned, and specialized document management software is used to store each document electronically and individually as an image file in a database.

Step 3: Individual documents are tagged or indexed to specific accounting transactions (such as journal or invoice IDs) or account codes (such as customer or vendor IDs) to create a link between the image file and a specific accounting system ID.

EVALUATION HEADS-UP

- Many client/server accounting packages integrate imaging either through a built-in ability to store images directly in the accounting database or by linking them to accounting data through use of a third-party product. Integration of third-party products is often through Microsoft OLE, providing another benefit of accounting packages that support this application integration technology.

- Legacy systems can be image-enabled through a separate imaging system that runs alongside the mainframe or minicomputer accounting system. Image indexing, using accounting transaction IDs or account codes, for example, lets operators toggle from the accounting screen to the image viewing screen and look up and view images linked to accounting data.

- Imaging adds a number of client and server hardware and software components to the system infrastructure, and these components add some cost and complexity to the acquisition and maintenance of the accounting system.

- As more accounting transactions are initiated electronically through EFT, EDI, or the Internet, the need to scan, store, and manipulate paper-based documents will disappear.

- Having images and documents viewable online adds significantly to the quality of accounting information. Online images reduce the time needed to answer many vendor and customer inquiries and make for a smoother auditing process because source documents are readily available for every transaction.

The resulting linked image file is stored on a separate image or document server. The image can now follow the accounting transaction wherever it goes. When users view a transaction on screen in the accounting system, there is usually a button or menu option that lets them view the linked document in a separate window.

Receiving Input Documents

A document that initiates an accounting transaction can be received electronically in many ways: via electronic data interchange (EDI), electronic funds transfer (EFT), e-mail, fax, or a transaction generated in an Internet browser applet. In all cases, a paper document may never have existed. Even the fax may have been sent directly from a word processing application. Both EDI and EFT depend on the use of specific electronic data formats and specific translation software to manage document flows between different systems. The translation software generates transactions directly from EDI or EFT data without human intervention. This may or may not be the case for the other electronic document delivery methods. E-mail, fax, and Internet transactions may be sent using specific formats that the appropriate accounting software module can parse to extract the information it needs to create a transaction. In most cases, however, a separate accounting transaction must be created and then linked back to the originating document. But there is never an envelope to open and discard or paper to copy and file, which alone can be a major time and resource saver.

Managing Electronic Output Documents

Accounting systems that support EDI or EFT can generate electronic documents that can be sent to other systems, usually belonging to suppliers or customers, to initiate transactions on those systems. Often the required EDI format is just one output option a system offers, in the purchase order entry process or the check payment process, for example. There may never be a need to create a paper document.

Many new client/server accounting systems are enhancing the function of electronic reporting. Instead of limiting you to printing and viewing reports, electronic reports also support data mining and electronic distribution. Once stored in electronic files, the report data can be analyzed by intelligent software agents looking for exceptions or unexpected trends in the data in order to generate electronically routed motifications, and reports or packs of reports can be disseminated electronically to organization-driven distribution lists via e-mail, fax, or the Internet.

Document Routing

Document routing is a means of delaying the point at which an accounting transaction is actually created by managing the document routing, usually some form of approval process, outside of the accounting system. You can use a groupware or electronic forms package to create documents such as purchase requisitions or sales orders and then route these electronic forms for review or approval via e-mail. Once the document has been through the approval process, it can become the basis for a new accounting transaction, such as a purchase order in the procurement system or a sales order in the order entry system. You can use document routing systems to front-end an accounting system, especially an older legacy system, with a new electronic approval process. This functionality does not require the accounting software to have workflow or document management capabilities; rather, a third-party software package manages the routing and pre-accounting workflows.

Document Decision Support

Imaging or document management software lets you easily annotate the document, add separate "sticky notes," or print the document. For accounting transactions, an electronic version of a source paper document provides the ultimate audit trail. Users can verify quantities, pricing, and approval signatures and check terms, conditions, and shipping data. As the document is routed, workflow participants can annotate the document directly or with sticky notes, providing reasons for approval or rejection and other information to help the viewer understand the document content. All users who can view the document have at their fingertips a rich set of information about the transaction to which it's attached. The information comes from a database that may be separate from the accounting database, so the performance of the accounting system can remain unaffected by the number and complexity of attached documents.

Approaches to Image-Enabling Accounting

There are at least two ways to image-enable an accounting system.

1. Provide seamless, transparent image-management functionality from within the accounting system to make it appear that imaging is an integral part of the accounting system.

2. Provide image management functionality as an adjunct to the accounting system, running alongside accounting software and linked by common tags or indexes such as a transaction or account ID.

Option 1 is the favorite of client/server accounting vendors who support Microsoft's Object Linking and Embedding (OLE) technology. Through OLE, third-party imaging software can be integrated with the accounting application, appearing to be just another piece of functionality delivered by the system. I discuss this option in more depth in the next section.

Option 2 is popular with users of legacy accounting systems, usually mainframe- or mini-based software unsuitable for integration with imaging software or too old for the effort integration would require to be worthwhile. In this case, imaging typically runs on a platform separate from the accounting system, such as a Unix server-based system or a local area network (LAN). Images and documents are scanned, stored, and manipulated using terminals or PCs that run the imaging, document management, or workflow software. The link with the accounting system is achieved only when the scanned images are manually tagged with a transaction or account ID to assist in retrieving the correct image for a particular inquiry.

Accounting users who need access to the scanned images replace their mainframe or minicomputer terminals with PCs. The PCs run terminal emulation to connect to the accounting system and simultaneously run the imaging, viewing, and manipulation software. Users can toggle between the two systems to manage inquiries — for example, to look up the account balance of a vendor in the accounting system and check the invoices that have been imaged and linked to that vendor ID in the imaging system. Although not the ideal, application toggling is simple and easy to implement and enables legacy systems to take advantage of online image and document management.

Image Enabling Through Microsoft OLE

You may wonder what relevance Microsoft's OLE technology has to accounting systems. Document management is one area where OLE can add real value to the management of certain accounting processes. If your accounting application is OLE-compliant (not all are), it can interact with other OLE-compliant applications. OLE provides a standard way for applications — in this case an OLE-compliant image or document management application and an OLE-compliant accounting application — to collaborate with each other.

Although the mechanics are quite complex, what OLE does in this case is let the accounting application call up a document management application so users can view or manipulate documents from within the accounting module. Users actually can use a separate application within its own pop-up window and be returned to the parent accounting function when they've finished what they're doing in the child application. The documents users see and manipulate are not usually stored in the accounting system's own database, but in a separate database. This situation has a number of benefits:

- Textual and numeric accounting data is not mixed up with image data.

- The accounting database is not enlarged by storage-hungry image data.

- The image database is optimized for handling image storage and retrieval.

- The image server can be scaled up separately from the accounting database server.

OLE makes it appear as though the document management functionality is built in to your accounting system and lets you enable the same functionality across a whole suite of accounting modules, from financials to manufacturing.

One popular product for image-enabling accounting software using Microsoft OLE is Watermark Enterprise, an image management product from Watermark Software, Inc., a subsidiary of FileNet Corporation. Like other imaging vendors, Watermark provides server software optimized for storing, indexing, and managing images and client software for scanning, viewing, and manipulating them. You can manage the images on the server using a hierarchical image browser to navigate and view images and to view and open folders of related documents. Client software lets users view and manipulate images retrieved from the image server. These viewers often provide a customizable toolbar that delivers the wide range of document management functionality summarized in Table 7.1. Any document that can be loaded into the viewer can be manipulated using the toolbar functions. The viewer is also e-mail-enabled, so users can e-mail any image they can view to other users of the e-mail system.

Table 7.1: Typical Image and Document Management Functionality

Document Annotation Tools	Image Manipulation Tools
Add text	View length of page
Add sticky note	View width of page
Add other OLE objects	View thumbnails
Add sound	Pan or zoom image
Use redliner pen	Magnify image
Use highlighter pen	Rotate image
Use arrow marker	Invert black and white
Erase text	Image deskew (straightening)

As an example of how image-enabled accounting works using OLE, the screens below show FlexiInternational's client/server accounting package FlexiPayables. Similar concepts apply to the products of other client/server accounting vendors that have image-enabled their software.

To link images to an accounting transaction, the user first sets a flag in FlexiInternational's system settings option to tell the system to switch on image-enabling. Setting the flag ensures that the buttons and menu options related to image-enabling are made visible and ready to use. After the flag is set, View, Scan, and Reject buttons for managing linked images appear on screens within various functions. For example, if the user is processing an accounts payable (AP) voucher, a Scan button appears on the voucher entry form, as in Figure 7.3. If a desktop scanner is linked to the user's PC, he or she can click the button to scan in the original vendor invoice. In this case, the imaging software actually manages the scanning, stores the document on an image server, and tags it with the AP voucher ID to create a link between the accounting transaction and the invoice document. After scanning the document, the user can check the image by clicking the View button. When the voucher entry process is completed, the voucher is saved and ready for electronic approval.

Figure 7.3: An Invoice Entry Screen with Imaging Enabled

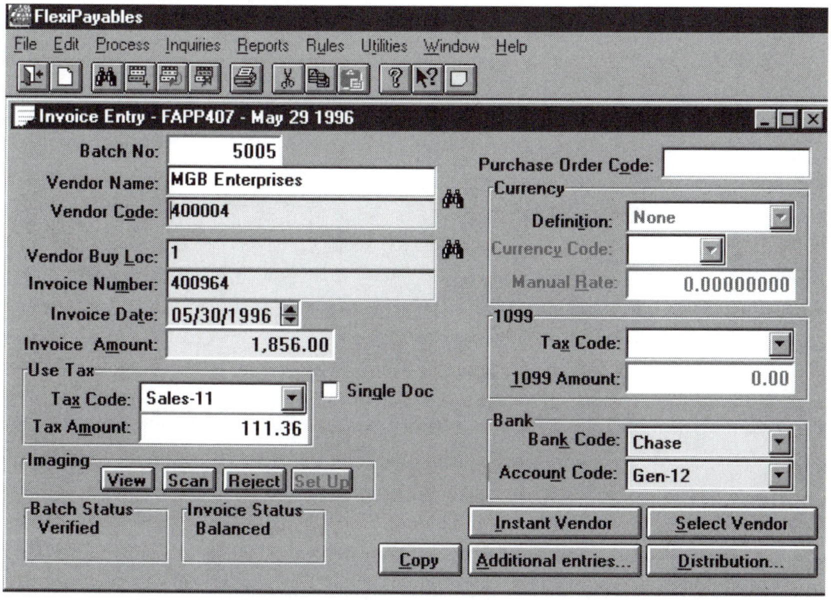

© FlexiInternational Software 1996

ScreenHelp: This screen shows the accounts payable voucher entry form in FlexiInternational's FlexiPayables after the system imaging flag has been enabled. Clicking the View, Scan, or Reject button sets imaging software in action and lets the user view a linked image, scan an image to link to the voucher being entered, or reject an image if the scan doesn't work correctly.

An approver can review invoices on screen using a list displaying unapproved vouchers. After drilling down to the voucher details form, the approver can click the View button to display the linked document, as shown in Figure 7.4. The document can then be manipulated in any way the image viewer toolbar permits. For example, the approver can annotate the document, highlight item lines, or add notes. The imaging software stores the annotations, highlights, and notes so they follow the document wherever it goes.

Figure 7.4: An Invoice Inquiry with Document Display

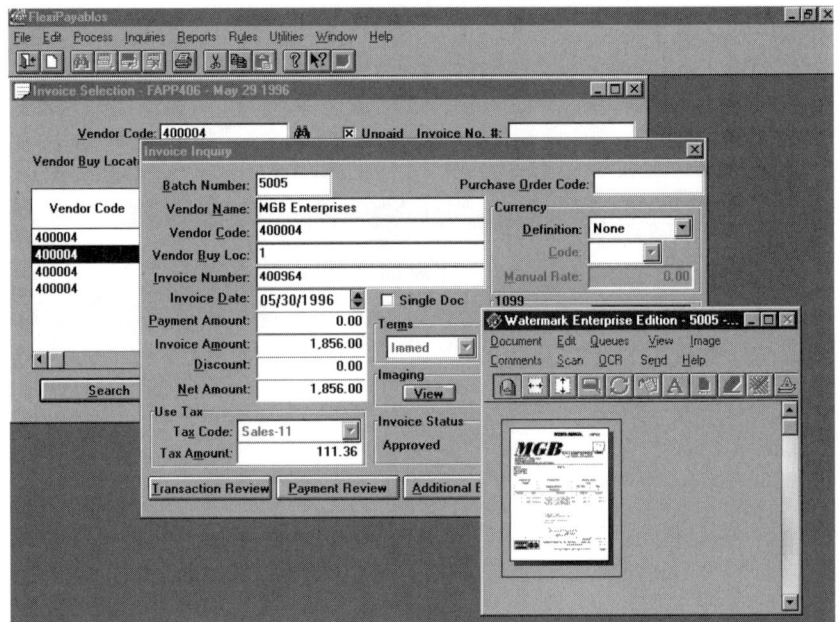

© FlexiInternational Software 1996

ScreenHelp: This screen shows how a user can drill down to the voucher image in FlexiInternational's FlexiPayables module. The invoice selection screen at top left shows a query against a vendor for invoices posted to the vendor account. On the invoice inquiry screen, the user can view the invoice voucher entry header and detail data. Clicking the View button on this screen displays the linked invoice document in a window controlled by the imaging software.

When the View button is clicked, OLE technology actually loads the image viewer in a separate window to display the document tagged to the current AP invoice. When the user is done viewing and manipulating the document, he or she can close the document viewer and continue the approval process. The process is similar when the user adds, views, and manipulates images in other accounting modules.

For low-volume document management with accounting, desktop scanners may be an efficient and cost-effective way to go. But in a high-volume transaction processing environment, this approach isn't practical. Instead, documents are scanned into the image server using specially designed high-volume scanners. Users then use imaging software to recall documents individually from the image server and tag them to accounting transactions or records. A user enters the necessary data, such as an invoice number, an inventory item, or a fixed-asset code, on a simple pop-up form. The tagging process can also record other audit data, such as who tagged the document and when. The tagging data is stored separately from the document, often in a relational database such as Microsoft SQL Server. Once the links are created, the tagged documents can be viewed and manipulated as described earlier.

Conclusion

As with e-mail, imaging should become available in all accounting systems within the next few years. Look for images to be available wherever they make sense throughout accounting suites and for the ability to associate multiple, user-navigable images with a transaction or item record. Systems will use the right mouse button to pop up a context-sensitive menu for accounting transactions or item records on which one option may be to display and manipulate a linked image. You'll also be able to use the image systems to create routing links between systems. An image of a purchase order, for example, may let you click the image or a button to display information stored in another image, such as the terms and conditions for the purchase order or corporate guidelines on purchasing specific goods or services.

QUESTIONS FOR VENDORS

1. Is the accounting software Microsoft OLE-compliant? If it is, then it's more likely the product can be image-enabled in the future, if imaging is not already provided.

2. If imaging is supported, what third-party package provides the functionality? Imaging functionality provided by a well-known imaging vendor is likely to be a better bet than a "one-off" imaging solution provided by a vendor reseller or consulting partner.

3. Does the imaging software support popular scanning standards such as TWAIN?

4. Does the imaging functionality extend beyond simply linking the image to accounting transactions and viewing the linked image? Can the image be annotated, linked to other images, or e-mailed to other users in response to image-related inquiries?

5. If desktop scanning and linking of images is impractical, does the imaging software support high-volume scanners and a fast, user-friendly way of collecting data for or automating the tagging process?

6. Is the imaging data stored with the accounting data or on a separate imaging server? A separate server is better for performance and scalability reasons.

Chapter 8

Internet
Accounting

Internet Accounting 101

Internet accounting is the integration of accounting applications with the Internet. Internet accounting lets users access accounting systems from any Internet connection through platform-independent browsers.

Although largely unheard of before 1995, Internet accounting already looms large in the future strategy of many accounting software vendors. Internet accounting is set to have as big an impact on the design and deployment of accounting systems in the second half of the 1990s as client/server has had in the first half of the decade.

From a business perspective, the Internet and the secured, private networks called intranets have so far been used mainly for e-mail and for publishing corporate documents, sales collateral, catalogs, policies, and procedures. In other words, the Internet has been document-centric, used mainly to reach a wider audience through electronic publishing. However, Internet accounting is less about documents and publishing than it is about transaction processing and participation in transaction workflows.

Internet accounting currently focuses on expanding the functionality of accounting applications and extending their reach by Web-enabling them to

- interact with Internet e-mail and file transfer protocols for transporting data files and messages to users of the World Wide Web

- publish reports and the results of accounting-system queries as Hypertext Markup Language (HTML) pages so they can be stored on Internet Web servers and viewed using browser software

- make functionality available to distributed users as applets (self-contained applications that represent a small part of an accounting module) in desktop browser software or let remote users access applications from forms embedded in HTML documents

By enabling applications to generate messages and message attachments and send them across Internet-based e-mail systems, Internet accounting lets you disseminate reports electronically to a user or group of users. Similarly, the accounting system can generate notification alerts and automatically send them to managers' and executives' e-mail in-boxes. Reports and query results published as HTML pages and stored on Web servers are accessible across the Internet by any authorized user, not just by users of the accounting system that generated the information. Apart from extending the functionality of accounting applications, these new features extend the reach of the accounting system to users throughout an enterprise.

Deploying application functionality as forms within HTML pages or as downloadable applets is a more revolutionary step. In this scenario, Internet-connected users can use as little as a single screen of the accounting system's functionality to interact with the system — the ultimate in functional granularity. From an Internet browser, users could, for example, initiate a transaction, participate in a transaction workflow, run a query, or request a report — all without having any accounting software on their own PC. The functionality they need is delivered across the Internet in the form of a downloadable applet or as a form embedded in the HTML page displayed by their browser. This functionality could let Internet users, for example

- initiate transactions by entering a purchase requisition, a sales order, a timesheet, or time and expense data using standard Internet browser software

- participate in transactions by approving or checking the status of orders, invoices, payments, or journals as they progress through their workflows

- query transaction audit trails, account balances, aging balances, and so forth

- request that a report be run and e-mailed to their personal Internet in-box

Because Internet browser software is cheap or free, Internet accounting is a much more cost-effective way of integrating casual users into the accounting process than is loading accounting client software onto the PC of each of these users, who may number in the hundreds. Note that Internet accounting does not provide users with free access to accounting data, because access to accounting servers is still governed by user license

EVALUATION HEADS-UP

- Internet accounting will cause a sea change in accounting software. It will fundamentally change the way accounting software is designed, deployed, used, maintained, delivered, and priced. Expect a storm of activity in the Internet accounting area for the rest of the decade.

- Accounting vendors are responding to the Internet in a relatively consistent fashion. They are Web-enabling their software to allow reports to be published in HTML format for distribution across the Internet, transactions to be initiated from Web browsers, decision support inquiries to be made against accounting data from Web browsers, notifications or alerts to be sent via e-mail to Internet in-boxes, and software updates and support to be delivered across the Internet.

- Most of the activity in Internet accounting will consist of establishing corporate intranets to open up accounting data so as to better serve the needs of an extended enterprise of employees, customers, suppliers, and other external parties. These intranet systems will require extra servers and special security software, adding to the complexity and cost of client/server accounting systems.

- EDI providers will soon come under real pressure to adapt their business models and software to the Internet. Those that don't are unlikely to survive its impact. Business-to-business electronic commerce on the Internet will revolutionize the way accounting software is used to manage business processes such as fulfillment and procurement.

- Increasingly, software agents will be used across the Internet to automate and fundamentally change traditional accounting processes such as fulfillment, credit control, and inventory restocking. The result will be less human involvement in accounting transactions and more time for spotting opportunities and preventing loss or fraud.

restrictions. A 50-user license allows only 50 users to access the accounting system, whether those users are local or remote. What's more, the performance of a Web-enabled accounting system may be degraded by the increase in traffic that often accompanies Internet accounting.

Internet accounting also requires that servers be added to the network to handle the additional traffic, which adds to the acquisition and maintenance cost of the accounting system. Figure 8.1 is a diagram of a typical Internet accounting system showing the need for additional servers: a proxy server to manage security and establish a "firewall" between the Internet and corporate data and a generic Internet server to manage Internet-related files and connection services.

Figure 8.1: A Typical Internet Accounting System

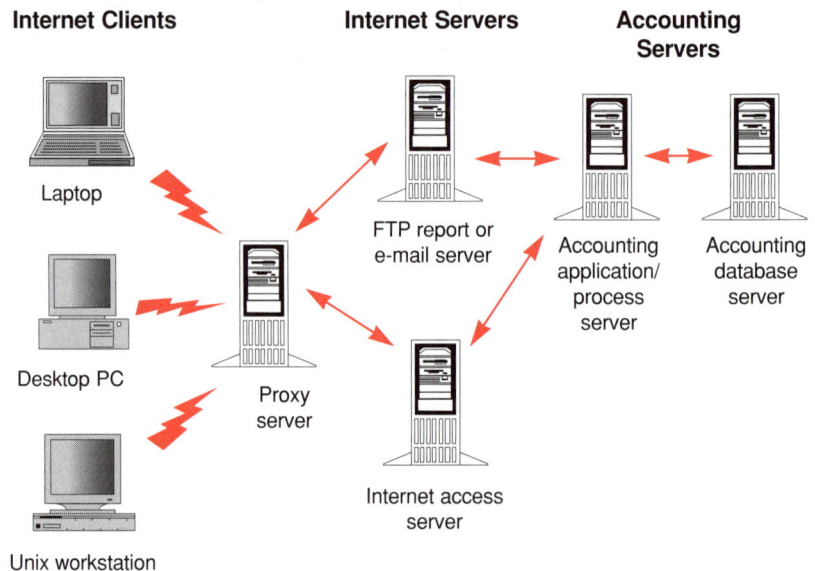

How Does Internet Accounting Work?

It's a relatively straightforward matter for vendors to publish reports or the results of queries as HTML pages. Most applications already save reports in a variety of file formats; HTML just requires adding another format to the options already available. Many client/server accounting systems are also e-mail-enabled, meaning they can interact with popular e-mail protocols such as Microsoft's Mail Applications Programming Interface. Again, the vendors of these systems can relatively quickly add support for Internet e-mail protocols so users can package messages and reports and send them to Internet e-mail addresses. In all these scenarios, data compression and encryption are important for fast, secure communications, so this functionality too must be added to every Web-enabled accounting system.

Deploying application functionality across the Internet, either as forms in HTML pages or as applets, is more complex. A purchase requisition function, such as the one shown in Figure 8.2, is likely to be a popular use for applets. To add this type of function, vendors must first build the HTML forms and pages and code the new applets, so you can expect them to start with the most obvious screens and functions and gradually add more over time. The application must also work closely with the Internet server and the proxy server and be able to maintain persistent connections with the accounting database for remotely connected users. Finally, the accounting application needs additional logic for displaying itself to a user who connects to the system over the Internet. Internet servers dedicated to managing accounting system users and traffic will become commonplace in all client/server accounting networks.

Figure 8.2: A Purchase Requisition Function on the Web

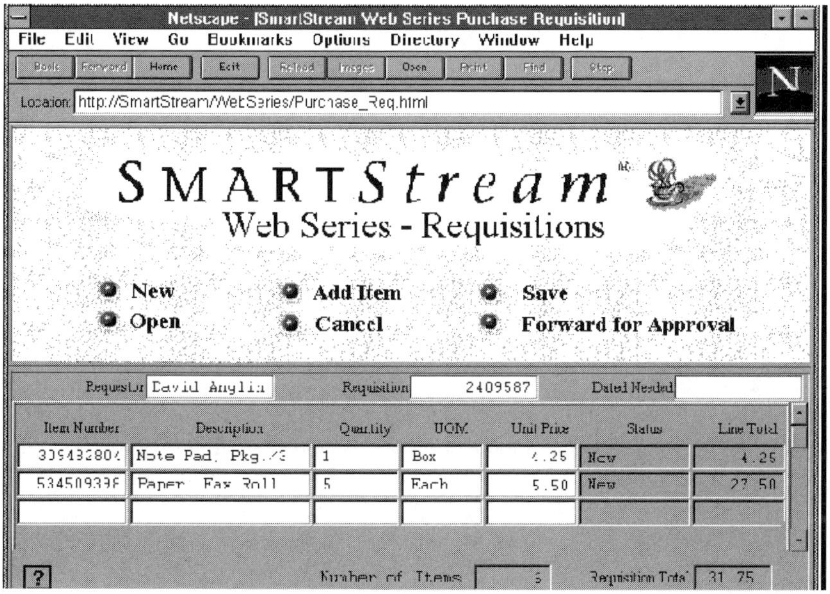

© Geac Computer Systems Inc. 1996

ScreenHelp: This screen shows an applet accessible to users of the Geac SmartStream Procurement application for requisitioning across the Web. The applet lets users create and submit new requisitions, view and edit existing requisitions, and route requisitions via SmartStream workflows to another user for approval.

Because Internet accounting is in its infancy, it's hard to say how some trickier problems will be handled. For example, for data entry across the

Internet, where should the validation services be located — on a remote server (just like a mainframe application) or within the local applet (slowing download times for the bloated applet)? As more casual users interact with the accounting system, data security may need to become much more granular to limit remote access more precisely. Also, as Internet access increases, should local users or remote users get priority? Or should access priority be based on who the users are or what they're doing? These are just some of the problems that need to be solved for Internet accounting to fulfill its promise.

Internet Electronic Commerce

Electronic commerce is the term used for electronic (paperless) management of consumer-to-business and business-to-business transactions. Much of businesses' interest in the Internet is in its potential as a revenue generator, as witnessed by the plethora of online storefronts and catalogs that let Internet users browse and purchase goods and services. Security concerns have triggered a great deal of debate about how to charge for purchases and collect money across the Internet, but the problems will undoubtedly be solved in the near future as standards emerge.

The increasing volume of goods and services ordered across the Internet is having a dramatic impact on the design of accounting modules such as sales order-entry systems. Sales order modules can receive orders electronically, eliminating paper, and can dispatch order-related documents, such as acknowledgments and shipping notes, via e-mail instead of by mail or fax. The new designs also change the workflow of sales order transactions and the role of order-entry clerks. An Internet order fulfillment process is fast and paperless and is managed by electronically enforcing business rules such as credit control policies. Order-entry clerks will become a dying breed, and sales order systems can be controlled by a single exception manager focused on handling such exceptions as orders for out-of-stock items or orders from customers that fail on-line credit checking procedures.

Although consumer-to-business electronic commerce receives most of the press, it's the development of business-to-business electronic commerce that will largely revolutionize how accounting systems are used. The reason is that business-to-business transactions must be end-to-end processes. For example, orders placed across the Internet must end up as transactions posted to order-entry modules, and these transactions must be managed across the Internet as they move through their lifecycle, so the entire fulfillment process can be completed electronically. The potential for such Internet transaction processing to change the way business is done is far greater than that of consumer-to-business electronic commerce. After all, placing an order across the Internet is only an incremental step forward from placing an order using a touch-tone phone. But managing the fulfillment process wholly electronically and possibly without human intervention is a much more significant step.

As business-to-business transaction workflows are reengineered to work across the Internet, they will change the role of electronic data interchange (EDI) as we know it. EDI is essentially a data mapping and business-to-business transaction management system. Business documents such as purchase orders are output to an EDI format file. Special EDI software then maps the individual data items of the purchase order on the customer accounting system to the corresponding items in the supplier's accounting system sales order. EDI software allows the heterogeneous accounting systems of business partners to behave as if they were homogenous. These paperless EDI transactions are managed across secure, privately managed, value-added networks. Today, EDI is synonymous with electronic commerce.

The combination of the Internet and more sophisticated workflow management software already has the potential to make EDI software obsolete. Consider the following scenario, which is not far off. An accounting system's purchasing module has a software agent that routinely checks for new approved purchase orders stored to the database. When it finds one, the agent packages the order data into an e-mail message, compresses and encrypts it, and e-mails it to the supplier's Internet address. This package may use existing EDI transaction formats or a new format such as the Open Application Group's (OAG) business document format. The supplier's sales order module has a similar agent that routinely checks the Internet mail in-box for special purchase order messages. When it spots one, it reads and parses it and then maps the data to an unapproved sales order transaction, which it inserts into the sales order database. When the order is approved, the agent automatically e-mails an order acknowledgment to the customer.

This process requires no EDI software (although standard EDI transaction formats may be used), and it lets expensive value-added networks be replaced by extended-enterprise intranets. This type of intranet is already being dubbed an "extranet" because it is an intranet that is not limited to a single corporation, but shared by a number of participating "external" business partners. Major EDI users such as General Electric Corporation are already moving their EDI traffic to the Internet, and every EDI software provider is examining its business model in light of the Internet's increasing influence.

Internet Accounting Players

Most leading accounting software vendors in the United States have already announced Internet accounting strategies, and some have even delivered working products. Almost every Online Analytical Processing (OLAP) vendor has announced or delivered a product for using OLAP software across the Internet for decision support. Document and workflow management vendors have announced or delivered products for accessing their document servers or managing workflows across the Internet. Microsoft, Netscape, and others are battling for the Internet server piece of the pie, and a host of new electronic commerce vendors have sprung up to deliver functionality such as data compression and encryption or secure credit card authorization processing. New Internet accounting applets are being built using SunSoft's Java, Microsoft's OLE/ActiveX, and OneWave's OpenScape tools. New concepts, such as software "wallets" that are used by participants in Internet transactions to identify themselves and define their preferred payment and shipping terms, will also become a key part of the Internet accounting process. All this activity ensures that Internet accounting will be one of the most dynamic software markets for the rest of this decade.

Intranet Accounting

The Internet has the potential to allow virtually everyone to access an accounting system from their desktop, but clearly, few businesses will open their accounting systems to that extent. Consequently, most of the activity in Internet accounting is expected to take place on corporate intranets — private networks connected to the Internet for data access and communications and protected by firewalls or proxy servers, which provide security for the accounting system by preventing unauthorized access to the accounting application logic or data. Although the security of data on the Internet, even when the data is managed on intranets, remains in doubt, new encryption software and new techniques, such as "tunneling" the data directly between two network addresses to make it difficult to interrupt, make the Internet more secure every month.

Because an intranet functions like a private network riding on the back of the worldwide Internet communications network, it can work as an extended-enterprise transaction-management system — a role previously monopolized by EDI systems. An extended-enterprise intranet (or extranet) can significantly expand the functionality and extend the reach of an accounting system when dealing with a variety of stakeholders in the business. For instance, suppliers can

- monitor their customers' inventory levels daily to help anticipate demand and plan production schedules

- e-mail shipping notes that could automatically create skeleton receipt transactions for completion and approval in the customer's goods receipt module

- receive payment by electronic funds transfer when the customer receives the goods without sending an invoice and waiting for it to be matched and approve

- automatically generate order acknowledgments, out-of-stock notifications, goods-shipping notes, and invoices and deliver them to customers by e-mail

- automatically e-mail special offer and pricing information to selected customers when the supplier enters the offers and prices in its pricing module

- have a credit-control software agent automatically e-mail a dunning letter as soon as a customer's outstanding debt reaches a certain threshold

Customers can

- place orders via the Internet and track their status through inquiries generated from standard Internet browsers

- e-mail a purchase order automatically to a supplier when an inventory item reaches its reorder point and triggers a supplier-notification software agent

Employees can access such human resource and company-related information from their Internet browsers as

- cafeteria benefit plans and balances on pensions or 401K plans

- operational or performance statistics for production teams

- copies of their W-2 or other payroll- or tax-related documents

And interested partners outside the company can access such information as

- company financials and performance reports

- rankings of key customers or suppliers

- revenue statistics for specific product lines or markets

When an accounting system is a component of a corporate extranet in these ways, it becomes nothing less than a complete management information and transaction management "console" for the business. Figure 8.3 illustrates some of the capabilities possible with a synergistic Internet/intranet structure; such a console is about as far as you can get from the traditional "information island" mentality that has dogged accounting systems in the past. Again, it's important to acknowledge that systems like this will fundamentally change the role of accounting clerks and managers, whose focus will change as systems are transformed from paper-driven, manual-entry processes to electronic transaction processes managed by software agents. Managers' focus will shift to analysis, and clerical roles will all but disappear — all that will remain are business analysts and exception handlers. This trend foretells the greatest shake-up of the accounting department since the introduction of computerized accounting systems.

Figure 8.3: A Web-Based Decision Support Console

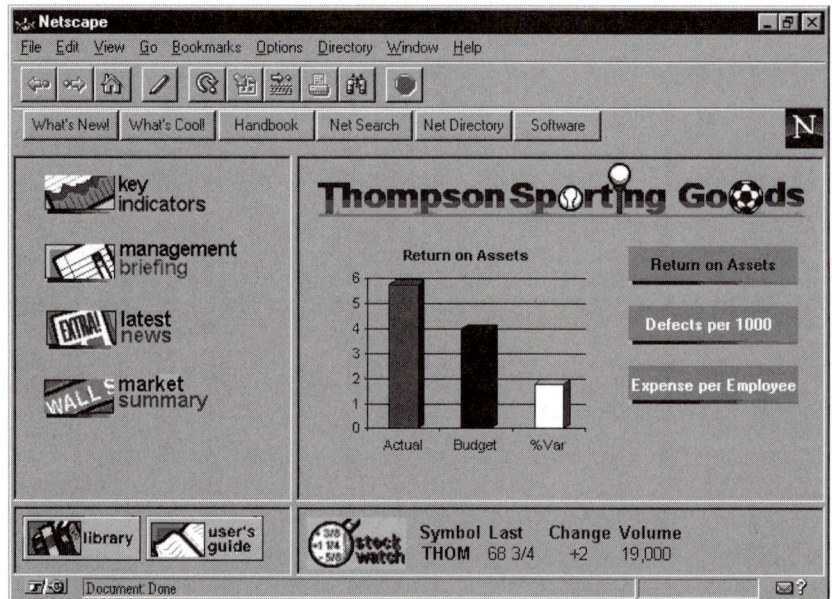

© Hyperion Software 1996

ScreenHelp: This screen shows a browser-based decision support console for Internet/intranet users of Hyperion Software applications. The console combines management information culled from various Hyperion accounting and reporting applications with data sourced from external service providers such as newsfeeds and other topical analysis.

Self-Service Accounting

Self-service accounting is another new direction being facilitated by the Internet for accounting software. Self-service accounting provides data owners more direct control over their data via the use of applets managed with Internet browsers. Self-service accounting can let

- employees maintain and query their benefit plans in human resource modules

- customers maintain and query their accounts in accounts receivable modules

- vendors maintain and query their accounts in accounts payable modules

- manufacturers query their item balances in inventory systems to assist with production scheduling

Figure 8.4 (page 96) shows a sample screen from a self-service customer account management module.

Self-service accounting is likely to be an intranet or extranet rather than Internet accounting function because of the persistent security concerns about the Internet. To implement self-service accounting, an enterprise must have applications with highly granular security capable of granting extended-enterprise partners very limited and specific access to only the data they "own."

ScreenHelp: The screen on page 96 shows Web Insight, a browser-based self-service account management form for intranet users of Software 2000's Infinium Financials applications. With Web Insight, customers can query their own account balances, history, and credit information and manage their billing and shipping address data. They can also get answers to their own invoice inquiries and correct their address.

Software Agents

Software agents will be crucial to the success of Internet and intranet ac-counting. A software agent is a piece of code that responds to a specific event, usually an event in a database, by carrying out a specific action. Located on either client or server computers, software agents simply wait in the background until "awakened" by the event they are designed to moni-tor. After completing their task, they return to their dormant state until the event occurs again, when the cycle repeats itself.

Software agents are "trained" by defining their behavior in maintenance forms — telling them the events they are to monitor, the rules they must fol-low to respond to those events, and the action to take in response to specif-ic combinations of success or failure of rule criteria. Software agents can monitor and respond to such events as

- a user attempting to log in or out of a system

- a database row being inserted, updated, or deleted

- a value exceeding a predefined threshold

- a specific report or process being run

- a new e-mail or workflow item being received by a user

Figure 8.4: A Self-Service Customer Account Management Application

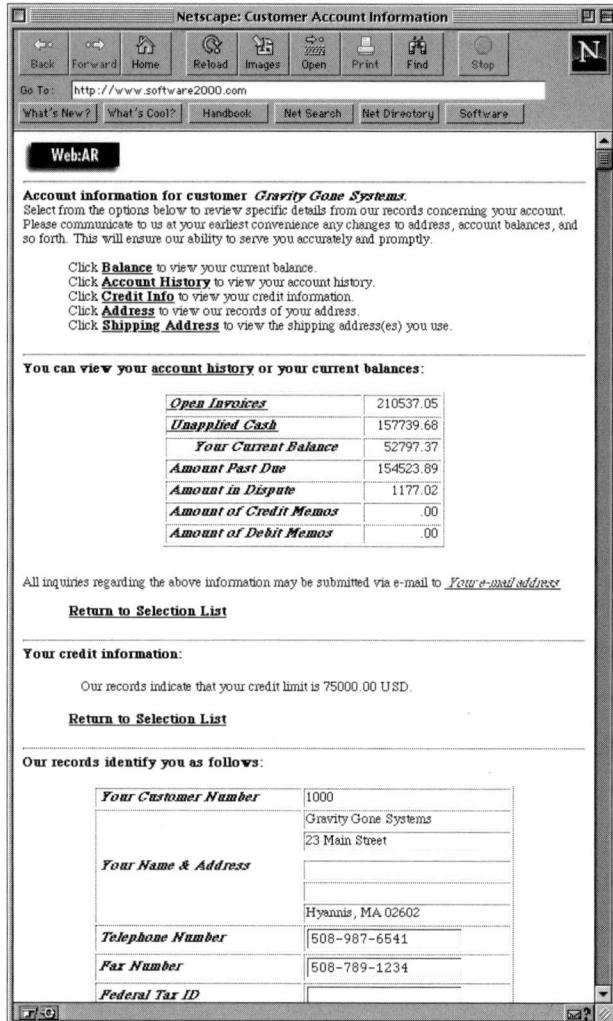

© Software 2000 Inc. 1996

An event can be tied to a highly complex set of rules that let the software agent apply some degree of intelligence to the event and decide what action to take. For example, in response to the actions listed above, a software agent might

- prevent a user from logging in or out of a system

- create an audit trail of database row inserts, updates, and deletions

- send an e-mail alert to a manager to warn of an exception condition

- initiate a series of actions after a report or process is run

- pop up a message to a user to tell him or her that new mail or work has arrived

A software agent can provide great deal of functionality for automating processes and providing a wholly exception-driven accounting environment. The Internet is an ideal routing mechanism for the software agent to use and for the deployment of software-agent-based applets that can function on Internet Web servers or on Internet clients using standard browser software.

Store-and-Forward Messaging

One problem with managing accounting transactions across the Internet is maintaining a persistent state for the transaction while users routinely connect and disconnect from accounting servers or when connections fail for any reason. Store-and-forward messaging is one way that transaction integrity can be maintained. Store-and-forward messaging stores messages until conditions are right to forward them to the recipients. For example, requests to approve invoices can be stacked in a server-based message store and then forwarded as soon as the approving user establishes a connection to the server. If a connection fails during message transfers, message stores can recognize the failure and retransmit the message when the connection is re-established. The approver's PC will also have its own local message store that behaves in exactly the same way to ensure that its approval messages are transmitted correctly back to the server.

New Accounting Paradigms

Software designed primarily to be run over the Internet, such as BHR Software's suite of applications written in Java, has the potential to resurrect an old concept with a new twist. When accounting applications were all mainframe-based and too expensive to acquire and maintain for all but the largest enterprises, organizations used timesharing service bureaus to run their accounting. The bureau stored the organization's accounting system software and all its data on a central mainframe computer that was shared among many clients. These clients paid a monthly fee for use of the remote computing resources, which, to use modern jargon, was simply a form of outsourcing.

Internet accounting offers the potential for a new type of "bureau" to be established, which might be called an "indenet" because it will be run by the independent software vendors themselves. In this scenario, accounting software resides on the vendor's internally maintained intranet, and each client stores its data on its own local intranet server. Application software costs would be much more manageable, being based on a monthly indenet membership fee, and involve virtually no maintenance burden for the member. In addition, with its data stored on its own intranet server, the member would still have control over and direct access to that data from other applications — something the old service bureaus could not offer. It remains to be seen whether this indenet business model is seen as viable by either software vendors or users.

Conclusion

By acting as a catalyst for accounting suites based on application frameworks and business objects, Internet accounting will change the way accounting systems are designed and deployed. For accounting system managers, the result will be more cost-effective and flexible systems that can be assembled around the needs of individual business units and business processes. By making redundant the concept of the accounting module and replacing it with the idea of a functional "granule," Internet accounting will also change the way software is priced. Instead of every user requiring a complete accounting module, certain casual users may need only a few forms or applets they can use from their Internet browser software. Internet accounting will also transform the EDI market and is likely to dramatically change the design and use of EDI software as we know it within five years.

Internet accounting will electronically reengineer basic business processes such as orders and payments, making them largely paperless, automated, and software-agent-driven. Receivables and payables departments will look much different than they do today both in terms of number of staff members and the role that staff plays. Most Internet accounting will in fact be deployed on intranets. Incorporating intranets will encourage the development of true extended-enterprise accounting systems, where accounting databases are opened up in various degrees to employees, customers, suppliers, outsourcers, and other enterprise stakeholders.

Internet accounting will allow a new injection of innovation into accounting systems and introduce a range of Internet accounting add-ons from a new generation of software startups. New business models, such as the indenet discussed above, will also undoubtedly emerge as both vendors and users recognize the potential of the Internet for accounting. Internet accounting promises an interesting future for accounting system managers and users in terms of their day-to-day roles and demonstrates how technology can be leveraged to catalyze the transformation of traditional business processes.

QUESTIONS FOR VENDORS

1. Does the accounting application allow reports and query results to be published in HTML formats and loaded on Internet/intranet servers for "pull-down" report distribution from Internet browsers?

2. Does the accounting system provide simple HTML forms for querying or updating information stored in the accounting database, letting you deliver self-service accounting to employees, customers, and vendors?

3. Can specific pieces of the accounting system's functionality, such as purchase requisitioning or budget submission, be utilized through applets that can be run from within Internet browsers?

4. Does the accounting vendor have a clearly articulated Internet accounting strategy? Is the vendor partnering with third parties that can supply payment transfer or storefront methodologies for delivering electronic commerce solutions?

Chapter 9

Workflow Accounting

Workflow 101

Workflow accounting is a business-process-driven approach to accounting software design and use that is largely a reflection of the current emphasis on corporate downsizing and reengineering, coupled with customer-centric and product-quality initiatives aimed at achieving "best-practice" results. These strategic business objectives have focused attention on the tactical business processes that define an enterprise's day-to-day operations, particularly those processes that are mission-critical and have the potential to add the greatest value to the bottom line. At least three types of workflow enabling are being provided in client/server accounting software:

- message-based workflows
- form-based workflows
- transaction-based workflows

Message-Based Workflows

Perhaps the simplest form of workflow accounting being implemented is message-based workflow. By using e-mail systems such as Microsoft Mail or Lotus cc:Mail, accounting systems can automatically generate messages or create attachments to be sent with messages and route them to any e-mail-connected user. This type of workflow is often initiated automatically through the intervention of software agents or triggers working directly against the accounting database.

The types of messages that can be sent include

- informational messages (e.g., "Your report is printed on Printer 1 and ready for collection")

- reminders (e.g., "Your in-box now has 5 journals awaiting approval")

- alert messages (e.g., "You have overrun the departmental budget for travel this period")

Attachments that can be sent might be

- audit trails, financial statements, and other reports in electronic formats

- charts or worksheets resulting from running queries against the accounting data

- "briefing books" containing reports, charts, and other data for executive review

In message-based workflows, the messages and attachments usually let users view data rather than take a specific action, such as approving an invoice. Nevertheless, even such a simplistic form of workflow can minimize organizational paper flows, reduce human forgetfulness, alert managers to exception conditions, and extend the reach of the accounting system beyond the confines of the accounting department and into other operational areas of the business.

Form-Based Workflows

The next level of workflow sophistication is form- or document-based workflow. In this case, a business form or document, such as a purchase requisition, vendor invoice, or employee T&E worksheet, is captured electronically. The form may be received fully completed via fax or as an electronic file on the Internet, or it may be scanned as an image into the accounting system. In the latter case, the form is then filled in on screen and routed electronically via e-mail to participants in a workflow process such as a requisition-, invoice-, or expense-approval process. Depending on user-defined rules associated with the form, one or more workflow participants who perform a specific organizational role, such as a reviewer or approver, take action on the data.

Form-based workflow is often implemented by combining the accounting system with another package, such as an electronic forms manager (e.g., Symantec's FormFlow), a groupware package (e.g., Lotus Notes), or a worksheet (e.g., Microsoft Excel). The workflow may be initiated in the external package and eventually received into the accounting system as a valid accounting transaction. Here are some examples:

EVALUATION HEADS-UP

- If your business process needs are simple and consistent and map closely to the standard functions of packaged accounting systems, workflow accounting may be an unnecessary and complex addition to your accounting environment. Not everybody needs workflow functionality or can leverage it.

- If you're already using a workflow engine, such as FileNet's Visual WorkFlo or even Lotus Notes, look for accounting software that can leverage that workflow infrastructure by allowing accounting transactions to be transferred to and from the workflow software.

- Workflow depends a great deal on the availability of e-mail. If you don't have an e-mail system in place, implement one before moving forward with a major workflow initiative.

- Successful use of workflow for accounting applications depends on your having a good grasp of the business processes you are running or want to run. Defining and flowcharting the business processes first will let you more effectively leverage workflow functionality when you deploy it.

- Don't limit your business process reengineering to transaction- or document-based processes such as procurement or fulfillment. Use the opportunity to examine other processes, such as month-end reporting, credit control, and decision support, to see how workflow principles can also be applied to them.

- Workflow-architected accounting software can give you the means to completely redesign the way you do business, the way your accounting systems are deployed, and the way your users interact with the system. Prepare everyone in your organization, from the top down, for the paradigm shift that can result from a completely different way of using your accounting systems.

- Look for maturity in the workflow functionality offered by accounting packages. Signs include integrated graphical process-mapping tools, universal user in-boxes and to-do lists for managing workflows at the desktop, secure auditing of workflows, collection and collation of workflow process statistics, and

Continued

EVALUATION HEADS-UP continued

graphical consoles for measuring workflow process efficiencies and for starting, stopping, and suspending workflows.

• If you are looking to implement extended workflow transactions that cross applications, look for table-to-table data mapping capabilities, support for middleware that facilitates moving data from one database format to another, and support for emerging cross-application workflow initiatives and interapplication data transfer from the Workflow Management Coalition or the Open Applications Group.

• A clerk creates a purchase requisition in Symantec FormFlow and routes it electronically, under FormFlow's control, for review and approval. Once the form is approved, FormFlow creates an accounting entry and inserts the entry into a special table the accounting system procurement module uses to create approved purchase order transactions.

• A saleswoman signs up a new customer in the field and enters the customer's address and contact data into a Lotus Notes database on her laptop. The next time the saleswoman connects to the corporate network, the local Notes server replicates the new customer data to the central Notes database and then routes it for the credit control manager's approval. Once the data is approved, Notes creates an accounting entry and inserts the entry into a special table the accounting system uses to add a new customer record to the accounts-receivable module.

• An employee fills in his T&E worksheet for the month and enters it into a special template in Microsoft Excel. The employee routes the completed worksheet to a manager for approval via Microsoft Mail. After the worksheet is approved, a Microsoft Visual Basic script embedded in the worksheet converts the T&E data into a balancing journal and saves it to the accounting system's unapproved-journal table for a supervisor's approval before it's posted to the system.

Alternatively, a transaction may be initiated in the accounting system, routed via the external package, and then received back into the accounting system. For example, a purchase requisition may be initiated in the accounting system, routed for approval through Lotus Notes, and then returned to the accounting system in its new state as an open purchase order.

Form-based workflow in accounting systems usually focuses on document-driven processes that require the documents to pass through various review, append, revision, and approval stages before the transaction is posted to the ledgers. Form-based workflow helps move businesses toward paperless accounting and the use of electronic in-boxes that can be managed on screen to deliver more efficient transaction throughput.

Transaction-Based Workflows

Although both message- and form-based workflows are process-based and require collaboration between an accounting system and another software package, neither is very complex to plan or implement. Both generally require a simple two- or three-step workflow that involves the collaboration of only two or three systems, such as accounting and e-mail or accounting, e-mail, and a forms package.

But in transaction-based workflows, the business process model demands that a number of applications or different parts of an organization work together to carry out a specific business process. A transaction-based workflow may route a transaction between many applications and across numerous organizational boundaries. The transaction lifecycle requires a more complex set of rule parameters to determine how and to whom the form is to be routed, the path the transaction is to take through the organization, and the people or departments to be involved. Known as "rules, routes, and roles," these parameters demand more functional depth in the software. The transaction is typically of long duration, meaning that it may take days or even weeks to complete, rather than the minutes or hours of a simpler workflow. And it may require multiple simultaneous workflows and interdependencies that make the workflow analogous to a Gantt-type project control chart. Such an extended transaction requires more sophisticated workflow management capabilities than does a form- or message-based workflow.

In the case of transaction-based workflows, the workflow software effectively becomes a transaction management and application choreographer, monitoring which transactions are in the workflows, their current state, and who is available to process them, and auditing the progress of the transactions through the workflow. Often the workflow essentially becomes an application within an application by combining functionality in unexpected ways, becoming in effect a "virtual" application that instantiates itself when the workflow starts and goes into hibernation when the workflow ends.

A purchasing-process workflow, for instance, may require the workflow engine to manage transactions that begin life in a purchase requisition form in Lotus Notes and then are processed by an accounting system's purchase, order, accounts payable, fixed asset and general ledger modules before the requisitioner is finally notified by e-mail that the requested items are available for pickup. This is a relatively straightforward workflow, but it nevertheless requires workflow technology to manage each transaction as it crosses up to seven applications; each application may be running on different computers using software supplied by several vendors and require the participation of many user workgroups that may also be geographically dispersed.

The complexity of transaction-based workflows means that the accounting software or its integrated workflow engine must offer very sophisticated functions, such as

- graphical process modelers for building and visualizing workflows on-screen

- software to track the state and transfer time of workflow transactions

- data collection forms for defining primary and alternate rules, routes, and roles

- workflow administration functions to initiate, audit, benchmark, redirect, suspend, and terminate workflows

- a way to manage receiving transactions from and passing them to other systems

- database middleware for translating workflow item data between different database formats

- knowledge of the network and applications environment for physically routing data around the organization

- sophisticated data import/export and mapping functions for passing data between applications

Transaction-based workflow systems let organizations map their accounting software much more closely to their business processes, rather than the other way around. These systems also let financial-systems managers use the accounting software's functionality to build customized applications that facilitate the day-to-day work of their users. In this way, workflow releases users from the restrictions of general ledgers, accounts payable, and the like by breaking down the traditional "hard" module structures into functional "granules" and lets users construct applications around their business processes by using workflow like a functional "glue."

The Universal In-Box

A big difference between workflow accounting software and traditional software is the way users view and interact with the accounting system. Workflow accounting makes for a proactive working environment by "pushing" tasks to users for them to act on or letting them "pull" applications onto their desktop to initiate a workflow. Tasks may be pushed to users by people above or below them in the organizational hierarchy. Users can pull applications from their PCs, from a local network server, or from a remote or Internet server; they can participate in workflows through their regular accounting software, desktop productivity applications, or Internet browser software; and they can push the results of their work up or down the organization.

In a workflow-accounting environment, users can "touch" applications only in places that make sense given their responsibilities in specific workflow processes. Most users don't need all the functionality of any specific software application, so they are often faced with a bewildering array of choices, many of which they don't understand and will never use. By simplifying what users see and how they can interact with applications, workflow accounting lets you deploy applications more efficiently and makes users more productive by helping them focus only on the functionality needed to complete their part of the task. In fact, many users of workflow-architected accounting systems need interact with the systems only through a universal in-box.

A universal in-box is simply an extension of an e-mail in-box. Universal in-boxes provide a single "console" from which users can manage all their electronic message traffic, which in addition to e-mail may include fax and voice mail, for example. Currently, universal in-boxes are just beginning to be integrated into workflow processing. Once they are in place, messages containing workflow items as attachments will let users launch a workflow to-do item from their in-box, process the item, and pass it on by sending it to the next participant's universal in-box. Similarly, to initiate tasks, users will employ special task messages to retrieve and launch the needed application from its local, remote, or Internet server. When accounting users have a universal in-box, they will no longer work with accounting "ledgers" and may not even be aware of conventional application boundaries. Instead, they will simply manage tasks and participate in workflow processes. The whole notion of discrete applications will disappear, and workflow will become the fundamental operating paradigm of all systems.

Why Is Workflow Important?

Workflow is an important functional addition to an accounting system for any size and type of business for four reasons. First, as software designers add workflow capabilities to the accounting system, their attention is focused on business processes rather than software features. The resulting software has more functional granularity and may offer out-of-the-box best-practice business models, so the application can be assembled closely to the way a specific business works or can encourage businesses to adapt their processes to best-practice models.

Second, workflow breaks down the barriers between accounting and other applications and between accounting users and other organizational users of accounting data. It does so by "pushing" information out to the users who need it and by crossing application boundaries to build process-based "virtual" applications.

Third, workflow accounting reduces paperflows, increases the visibility of information, reduces the need for human intervention, and helps an organization become exception-driven. It does all this by using electronic rather than paper distribution and by letting preprogrammed system rules and agents do jobs that humans too easily forget or ignore, or simply should not be doing.

Finally, making use of the workflow functionality offered by an accounting system demands that you have a clear handle on your business processes. Thus, an accounting system's workflow capability acts as a practical catalyst for the reengineering effort necessary to take full advantage of workflow functionality. Putting workflow technology to use is one way you can truly add value to your business processes and begin the task of converting your accounting users from transaction processors into knowledge workers.

Process Mapping and Monitoring

Visualizing workflows is an essential part of mapping workflow process steps to software functionality. Ideally, workflow accounting software supports visualization by letting you define a process flowchart on screen using a graphical tool, such as the one shown in Figure 9.1. You can then enrich each process step with workflow events and actions and map it to specific accounting data entry, inquiry screens, or nonvisual accounting processes such as validation, posting, or currency conversion. In this way you can build process models and compile them into virtual applications that reflect real-world business use rather than the software developer's anticipated use. Then it becomes practicable to try out the processes, reengineer them on the fly if necessary before deploying them, and accommodate ongoing process change.

ScreenHelp: This screen is from the Geac SmartStream workflow-architected suite of client/server applications. The screen shows the Workflow Workbench being used to graphically model a workflow process, in this case adding a new course to the training schedule. By double-clicking on the representation of each process step, the user can display tab folders for defining the behavior of the step (see Figure 9.2 for an example).

The primary potential benefits of using workflow software are improvements in process efficiency and reductions in process cost. To achieve these benefits, the workflow must be monitored for conditions and timings such as

- elapsed times — the total time taken to complete the workflow process

Figure 9.1: Visual Process Modeling

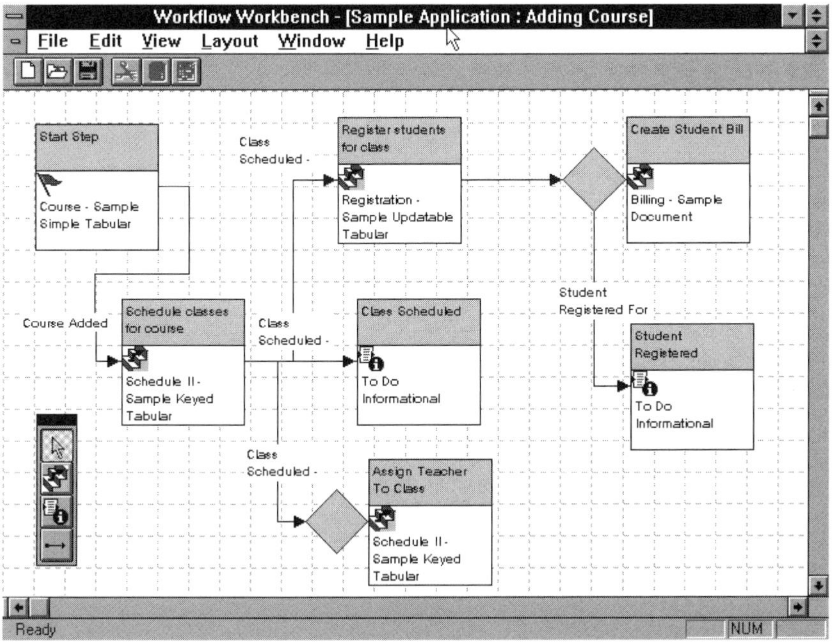

© Geac Computer Systems Inc. 1996

- activity times — the total time spent in each activity in the workflow
- transfer times — the total time taken to transfer between workflow activities
- wait times — the time the workflow is on hold pending a specific action

These sorts of timings, combined with monitoring process cost elements, can deliver much useful information about the efficiency of a process and enable meaningful benchmarks that let organizations compare the efficiency and cost of their processes with those of best-practice, cross-industry process leaders. Such data also helps determine the relative usability of the accounting application at the individual screen or activity level because it quickly reveals process bottlenecks and discrepancies among how individual users interact with the system. Ideally, this sort of information should be presented in the form of a workflow console that charts the statistics of individual processes and allows detailed drill-down and monitoring of specific workflow steps.

Rules, Routes, and Roles

Rules, routes and roles are an integral part of the workflow functionality offered by workflow-enabled accounting systems. Rules are conditions, usually based on an "if...then...else" construct, that determine how and to whom the form is routed. Examples of rules might be

Example 1

1. if the purchase requisition value is less than $100

2. then flag the requisition as approved and create a PO

3. else route requisition for approval to department manager

Example 2

1. if the vendor invoice has a due date less than five days from today

2. then alert the accounts payable manager to review invoice

3. else schedule for payment based on due date

Example 3

1. if the T&E form total causes the employee's period expense budget to be exceeded

2. then alert the department manager for review

3. else approve the T&E worksheet and create and post GL journal

Of course, rules can be a great deal more elaborate than this. They can even involve complex programming algorithms using artificial intelligence techniques such as fuzzy logic, heuristics, pattern-matching, and neural-network processing. Such complex rules are usually associated with special financial-services trading systems, however, and are seldom found in general accounting systems.

A route determines the path the workflow item (i.e., the electronic form) follows around the organization as it moves through its lifecycle. Routes are usually based on organizational hierarchies such as reporting trees, organization charts, or territory structures that are often already defined and maintained in the accounting software. In practice, a route usually comprises a series of links to user IDs in the accounting system or to external e-mail addresses. Rather than being links directly to individual users, however, routes are mainly links to roles, which in turn are linked to the users who participate in workflows.

Roles are workgroups of users who perform a discrete organizational function, such as entry clerks, supervisors, and managers. In practice, roles usually fall into one of the following types of categories according to the job they do:

- initiate transactions (e.g., invoice entry)
- review and question transactions (e.g., review of invoices)
- approve transactions (e.g., invoice approval)
- post transactions to the accounting database
- start, halt, suspend, or terminate workflows (e.g., system administrator)

Roles often have both a vertical and horizontal dimension to them. Many roles are hierarchical in that they have a manager role above them to which items are routed after the role participant has finished with them. Roles may also have alternate or surrogate roles at the same level of authority to handle overflows, logjams, or personnel shortages. Both of these dimensions are necessary for workflow items to be routed effectively and efficiently around the organization.

Figure 9.2: Defining Rules, Routes, and Roles in a Workflow Process

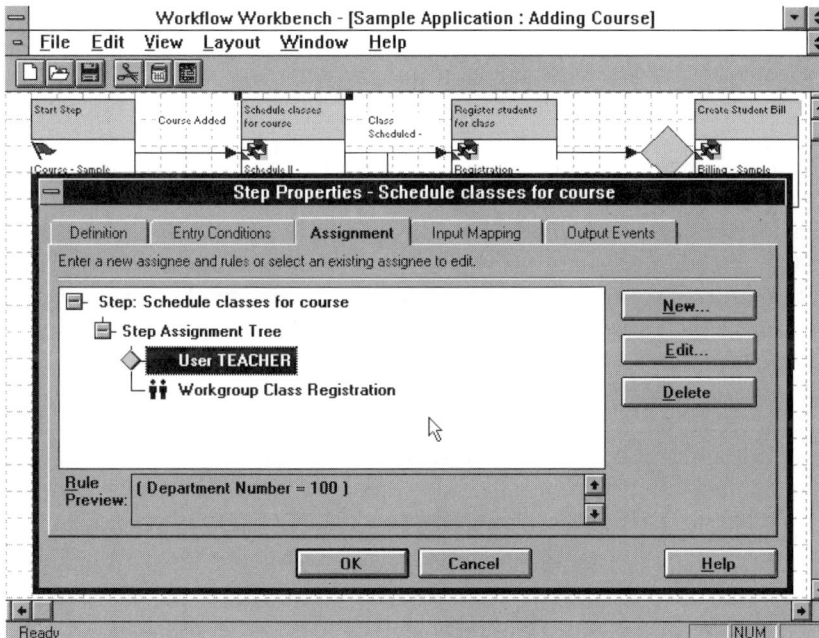

© Geac Computer Systems Inc. 1996

Because rules, routes, and roles must be maintained, they also make the administration of the accounting system more complex. Figure 9.2 shows how they are managed in one workflow-enabled accounting system.

ScreenHelp: This screen is from the Geac SmartStream workflow-architected suite of client/server applications. The screen shows the tab folder used to define workflow process activity steps and the events, rules, routes, and roles associated with them. On this particular screen a process step is being assigned a rule and route.

Activities, Events, and Actions

There's a difference between workflow-enabled and workflow-architected accounting software. Workflow-enabled software layers the workflow functions on top of the accounting software, usually in the form of integration with third-party workflow-capable packages such as Lotus Notes or FileNet WorkFlo. In this scenario, the workflow functionality is provided by the third-party application rather than being an integral part of the accounting software. Workflow enabling makes a lot of sense if you don't want to over-complicate your accounting software or if you already have a major investment in a workflow tool and want to leverage it by bringing your accounting system under its umbrella. You can also use this approach to provide workflow functions to older legacy applications that the vendor is unlikely to enhance with built-in workflow capabilities.

Workflow-architected software is a different animal. Such software is designed in a much more granular fashion, reflecting the fact that workflows are made up of a series of tasks or activities that can be combined in myriad ways depending on the business enactment of specific processes. Each task or activity may also be subject to an array of processing events that impact the way the activity is used within a given workflow. For example, if the activity is completed successfully, the workflow may progress in one direction, whereas it may progress in another if the activity fails. Or, if the workflow progresses according to plan, it may involve one set of participants; otherwise it may involve a completely different set of participants. Consequently, workflow activity events can be subject to multiple rules that may be nested so that certain rules are used only when their "parent" rule is triggered.

Workflow-architected software is designed to be deployed in a much more granular and customized fashion than traditional accounting software — a scenario that has lots of interesting possibilities. For example:

- Users may be able to define their own activities, events, and actions either as wholly new objects or as additions to those shipped with the system, allowing for flexible, "codeless" package customization.

- Systems can be assembled around workflows that map directly to a user's day-to-day job to provide a custom fit of technical functionality to business process.

- Functionality can be delivered in small chunks as standard screens or as applets or form scripts inside HTML pages across the Internet to accommodate mixed processing environments, including local and remote users, accounting and nonaccounting users, and simple or sophisticated data entry and inquiry.

Report Workflows

Workflow concepts are not applicable only to accounting transactions — they can also be applied very effectively to accounting reports. Report workflows assume that the production of the report is the beginning, rather than the end, of a business process. Report workflows also assume that reports are distributed electronically rather than on paper. The report itself may be static, allowing the user simply to navigate and view the data, or dynamic, allowing the user to drill down and uncover more detailed transactions. Dynamic reports are sometimes called "snapshots" because they contain both the logic for formatting and displaying the report and the transaction data that provides the basis for the report. Examples of report workflows include

- distributing reports electronically as attachments to e-mail messages

- distributing reports electronically as files to report servers that can be accessed remotely by mobile-computing users or downloaded across the Internet

- collating packs of reports into "container objects" called workbooks and then distributing the workbooks electronically

- archiving reports as images to image servers for long-term storage and retrieval

- saving reports as electronic files that can be automatically "mined" by workflow and notification engines, which can then disseminate exception alerts to managers or push work tasks to users based on the report data

- saving report structures and data as a snapshot database file that remote users can download for local analysis or append to their own personal report databases for ongoing analysis

Conclusion

Workflow technology is far from reaching its potential in accounting applications or elsewhere in business management software. Workflow has the potential to be used as the choreographer for managing

- a variety of accounting and other business processes

- interapplication interoperability — for passing data between manufacturing and financial software, for example

- resource allocation and transaction throughput, completion, and rollback

- interaction between business component objects and making these objects available to the applications that need them

- paper-free, proactive, exception-based notification and reporting systems

Neither the workflow vendors nor the accounting vendors have scratched the surface of this potential yet, so the workflow boom is far from over. In fact, it has barely begun.

QUESTIONS FOR VENDORS

1. If the accounting package provides workflow functionality, is it based on a third-party tool or is it built into the system functionality? If it's based on a third-party tool, which one?

2. What sort of workflow processing does the system support? Can you manage transaction approval workflows? Can you assemble functions around your own business processes? Or does workflow simply mean restricting users to seeing only the functions they use on a day-to-day basis?

3. If the package provides sophisticated workflow, does the functionality include visual workflow builders and consoles for monitoring the status of workflows?

4. Does the system audit workflows and keep timings so you can identify process efficiencies by analyzing the audit and timing data?

5. How are workflow exception events, such as interruptions or bottlenecks, handled? How does the system check the integrity of a workflow?

6. Can the system's workflows be extended across third-party vendor accounting systems or workflow engines? Does the workflow engine comply with any standards, such as those of the Workflow Management Coalition?

Chapter 10

Component
Accounting

Components 101

When you evaluate accounting software from a technology perspective, some of the questions that come immediately to mind are

- How scalable is the system — can it cope with my business growth?

- How well does the system perform under transaction load?

- How adaptable will the system be to changes in my business processes?

- Will I be forced to purchase software with functions that I don't need?

The answers to these questions depend a great deal on the granularity of the accounting software design and whether the software can be delivered in the form of components — individual discrete functional objects — rather than as complete modules.

Accounting software scalability, performance, and adaptability are all directly impacted by the software's granularity of design. Nature itself is the most granular of all systems, and also the most scalable and adaptable system we know. The reason is that nature builds objects from microscopic components, each of which performs a discrete functional task. Building software applications is really no different — granularity of design allows functional components to be isolated by task or role so that they can be assembled and deployed in different ways.

A good starting point for understanding how accounting software can be designed around a component model is to consider the four main layers of software functionality offered by most accounting systems:

Layer	Description
Presentation	What the user sees on the screen; the application user interface
Validation	The rules that ensure the data a user enters or requests is valid
Processes	The major system processes, such as posting and financial reporting
Database	The storage and management system for the accounting data

If you think about the construction of these layers, you'll see that there are finer levels of granularity within each one:

Layer	Lower Levels of Granularity
Presentation	Individual screen forms; individual data fields on those forms
Validation	Individual rules for specific forms or fields
Processes	Individual processes, such as posting journals or printing invoices
Database	Individual database tables and columns within those tables

Within each layer it's possible to isolate even finer levels of granular components and design them as discrete, functional tasks or objects within the application. Already you can see how easy it is conceptually to view an accounting system not as a set of modules, but as hundreds or even thousands of granules, each performing its own function within its specific functional tier. Building accounting systems with this level of functional granularity requires a completely different approach to software design, and it is only in the past few years that vendors have begun to use this object-oriented design methodology to build their applications.

Component Accounting Delivers Scalability

Component accounting systems can be deployed in a variety of ways. This flexibility is key to a system's ability to offer three types of business processing scalability:

EVALUATION HEADS-UP

- The granularity of the accounting software you select has a direct effect on how flexible it will be to buy, deploy, scale for growth, and adapt to your business processes. More granularity is better than less. Look for software you can purchase in components that deliver functionality at a lower level than the traditional GL, AR, and AP modules.

- Accounting applications that offer fine granularity will probably be able to better exploit the Internet and more effectively support mobile computing users because their functionality can be delivered in smaller, less resource-hungry pieces. As a result, you'll be able to expand the accounting domain and extend its reach to nonaccounting analysts and management users.

- Microsoft OLE 2.0-compliant accounting applications can already make some use of the component paradigm by integrating applications such as worksheets, charting, and imaging with accounting modules and functionality. In the future, OLE (and ActiveX) will likely offer compliant applications even more capabilities for sharing objects across both local and wide area networks.

- Software designed around the business object paradigm is a big jump from the modular "silo" design of traditional accounting software, in both its visual interface and the way it's used. Before you consider this type of software, make sure your people are ready to make the conceptual leap, as not everyone — particularly in the accounting department — is comfortable with paradigm shifts.

- No one should doubt that all accounting software eventually will be designed based on business frameworks and business-object components. Even if you aren't yet ready for this type of software, check that your vendor has a business-object vision and is staying abreast of the technology and its implications. True component-based accounting software has the potential to make even the new generation of client/server systems look like dinosaurs, and it will change completely the way accounting software is designed, deployed, and marketed.

- transaction scalability, for handling increasing transaction volumes

- connection scalability, for handling more system users

- functional scalability, for handling increasingly complex business processes

Consider the common deployments of the four functional accounting layers across hardware tiers, as shown in Figure 10.1. The more the functional layers can be isolated onto separate computers, the more scalable the solution can become. The scalability problem can largely be solved by throwing hardware resources at it — usually a more palatable and cost-effective option than changing accounting software. In most cases, this solution means adding more processors, memory, or disk resources to the accounting servers or simply adding more server boxes to scale up the resources available to the system overall. Compared to two-tier client/server deployments, you can scale a three-tier or n-tier system without upgrading all your client computers or changing what is usually the most expensive part of the system — the database server. Instead, you can simply add application or process servers, which can often be inexpensive PCs, to your network. The functional design granularity and ability of the software to be run as discrete components anywhere on the network are what facilitate three-tier and n-tier deployments.

Figure 10.1: Common Deployments of Functional Accounting Layers Across Platforms

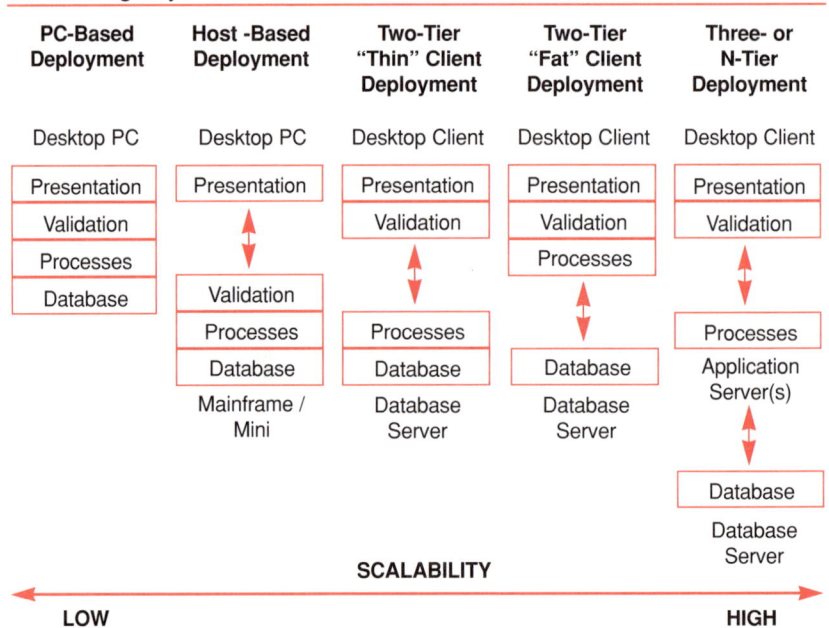

PC-Based Deployment	Host-Based Deployment	Two-Tier "Thin" Client Deployment	Two-Tier "Fat" Client Deployment	Three- or N-Tier Deployment
Desktop PC	Desktop PC	Desktop Client	Desktop Client	Desktop Client
Presentation	Presentation	Presentation	Presentation	Presentation
Validation		Validation	Validation	Validation
Processes			Processes	
Database	Validation	Processes		Processes
	Processes	Database	Database	Application Server(s)
	Database			
	Mainframe / Mini	Database Server	Database Server	Database
				Database Server

SCALABILITY

LOW ←————————————————————→ **HIGH**

Component Accounting Delivers Adaptability

In the past, the lowest level of granularity for most accounting software was a module, such as the general ledger or accounts payable module. Software users either had access to the module or not. In other words, the system was deployed as a set of vertical application modules, or "silos," rather than being designed to fit around common business processes.

Figure 10.2 compares the old modular software structure to more modern process- and component-based structures. A modular structure is no longer acceptable in today's process-driven world, where the typical accounting user is no longer just a transaction processor, but is on the way to becoming a task-oriented knowledge worker. Real-world functional tasks cross the modular boundaries of traditional accounting software design. So instead of being limited to using the accounts payable module, for example, a user may be able to manage a business process, such as procurement, from beginning to end by combining functionality from purchasing, general ledger, and accounts payable modules. To service this need for flexibility, component-based accounting systems let users assemble functionality around their business processes rather than the other way around. In component-based accounting, the terms general ledger and accounts payable are merely conveniences to indicate a collection of functional components that fulfill the role of the traditional GL or AP module.

Figure 10.2: Modular vs. Process vs. Component Applications

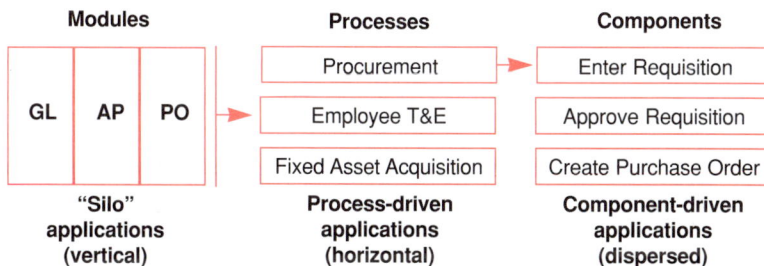

Modules	Processes	Components
	Procurement	Enter Requisition
GL AP PO	Employee T&E	Approve Requisition
	Fixed Asset Acquisition	Create Purchase Order
"Silo" applications (vertical)	Process-driven applications (horizontal)	Component-driven applications (dispersed)

A granular accounting system lets users easily assemble components around their own specialized tasks because the system's functions are small enough to allow this functional recomposition to be done according to the business's process needs. Workflow or process management software is responsible for choreographing the interaction between these functions. Note that this approach is not the old, security-driven mechanism of switching off menu options for specific users to hide functions from them. Instead, all system functions across all modules are exposed, so the user can mix and match functions according to the accounting process being automated and the user's responsibilities. This is the approach taken by

- Geac SmartStream's activity-based to-do lists
- LIBRA's IET user-defined workflow palettes
- Great Plains Software's Dynamics C/S+ pop-up Task Lists
- Ramco's Marshal drag and drop application configurator
- Lawson Insight's business process suites
- Navision Software's Navision Financials granule configurator

This ability of users to assemble software functions into task- or process-driven workflows illustrates granular accounting's adaptability. Another example of its adaptability is the use of a data dictionary layer in the software to act as metadata for the accounting database. Metadata, or data about data, provides information about a specific piece of data the accounting system uses and how the system uses it. Take an account code, for example. It may be just a string of letters and numbers when stored in the database, but the data dictionary provides other information about it, such as

- the maximum length of the code and how many segments it contains
- the label used to describe the account code on forms and reports
- messages and help text that can be displayed when the account code is used
- validation checks and rules used to ensure that entered account codes are meaningful

A central repository of information about all aspects of the accounting data at this level of detail benefits system users and managers in a number of ways. At a basic level, they can edit the dictionary data to reflect their own specific rules, terminology, or messages. Any changes they make to a data item in the data dictionary carry through the entire application, wherever the data item is used. So a single modification to a data item in the data dictionary will be reflected in every data entry form, report, and pop-up message window that uses that item.

An accounting system that has a data dictionary is often the best solution for an international business because the system can be deployed in a multilingual environment. The data dictionary can provide metadata in a number of languages; a user-defined log-in language determines which version the application shows a particular user. Thus an installation in Canada can easily accommodate a mix of English and French screen forms and reports, while the same system in a Swiss branch office can be run in English, French, and German. Software from European vendors such as Scala and Navision Software uses data dictionaries to deliver this type of linguistic adaptability.

Components Are Cost Effective

As accounting systems become more granular, the pricing will change and the software will become more cost effective to implement and upgrade. When you "roll your own" accounting system from individual components, you'll be able to buy only the pieces you need by choosing from a menu of components such as the one shown in Figure 10.3. You'll also be able to upgrade the system by adding small, easily deployed, relatively inexpensive pieces as your functional needs grow. Although we are some years away from a market that offers such components off the shelf, many vendors are examining this business model, and some are undoubtedly working on the first accounting systems built entirely from standalone components.

Figure 10.3: One Vendor's Menu of Components

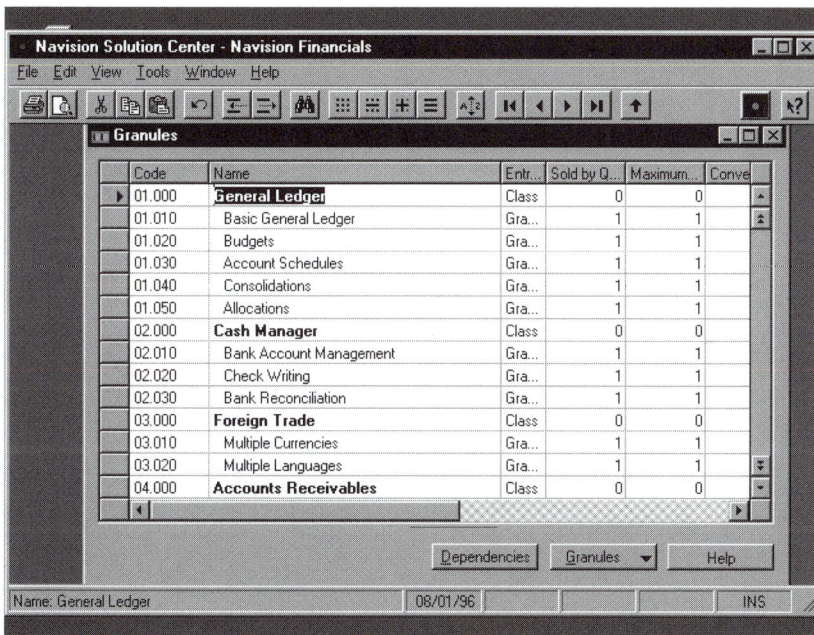

© Navision Software US Inc. 1996

ScreenHelp: This screen shows the application configurator from Navision Software's Navision Financials. Navision resellers can use this screen to assemble a customized version of Navision Financials from an extensive list of application components or "granules." The application is then generated specifically to the customer's requirements and can be easily upgraded by adding new granules using the same screen.

The vendors that have taken the most object-oriented approach to software design, such as Software 2000 and its Infinium Financials, are likely to be the first to deliver accounting components and introduce a component pricing model. Object-oriented software design drives vendors toward reusable components, termed business objects in the object-oriented world. The component approach also lets vendors focus their attention on the functionality offered by the most-used components and encourages them to continually broaden and deepen that functionality. The more service components in place in the application, the more quickly the vendor can deliver new functionality, because new components are based on tried and tested existing components and inherit much of their functionality. This object-oriented approach also means new components will have fewer bugs than the source code modifications of the past.

Frameworks

As component accounting matures, vendors will begin to design complete frameworks that model specific vertical businesses, such as financial services, retail, health care, oil and gas, and pharmaceuticals, as a set of interconnected software objects. These frameworks, consisting of interconnected business objects, represent real-world people, places, processes, and things, such as

People	Customers, vendors, employees, salespeople, contacts, clients
Places	Offices, warehouses, shipping depots, inventory racks and bins
Processes	Reviews, approvals, posting, currency conversion and revaluation
Things	Legal entities, orders, invoices, checks, timesheets, journals, projects

Frameworks of business objects will be usable as-is, with minor modifications, or as the starting point for the construction of specialized vertical-market accounting systems built on to the framework's functionality. The framework becomes the "source code" for these systems, but at a much higher level of abstraction than conventional program source code. When customization is required, the customization process is also undertaken at a much higher level than is source-code customization and is easier to manage and less prone to error.

It will be practicable for internal or third-party business analysts to customize frameworks to suit the needs of specific businesses. Corporations wishing to extend the framework's functionality, however, will continue to need programming specialists for the foreseeable future, until the market for third-party business objects matures. Major market players such as IBM, Microsoft, and SAP are working on framework projects and expect to deliver usable business frameworks, possibly for the banking or insurance business, consisting of specific line-of-business objects, such as currency conversion or production control scheduling, in 1997.

Business Objects

Each business object is designed to be a truly independent functional component that can be deployed in various combinations to create customized frameworks. The eventual goal of object-oriented software design is to let customers purchase business objects from multiple vendors and assemble them on the fly into a homogeneous system. This vision holds the promise that customers will be able to assemble the ultimate best-of-class accounting application for their enterprise by combining best-practice business objects sourced from different vendors.

The real problem with business objects today is not so much that they can't be built, but that they are difficult to deploy. Business object standards are in their infancy, and the infrastructures for managing business objects are immature. However, accounting systems built on a form of business object model exist today — Software 2000's Infinium Financials (Figure 10.4) is one of these. Other vendors, such as System Software Associates and Marcam, are also focusing on this type of design model for their new generation of accounting systems.

Figure 10.4: The Main Business-Object Interface for Infinium Financials

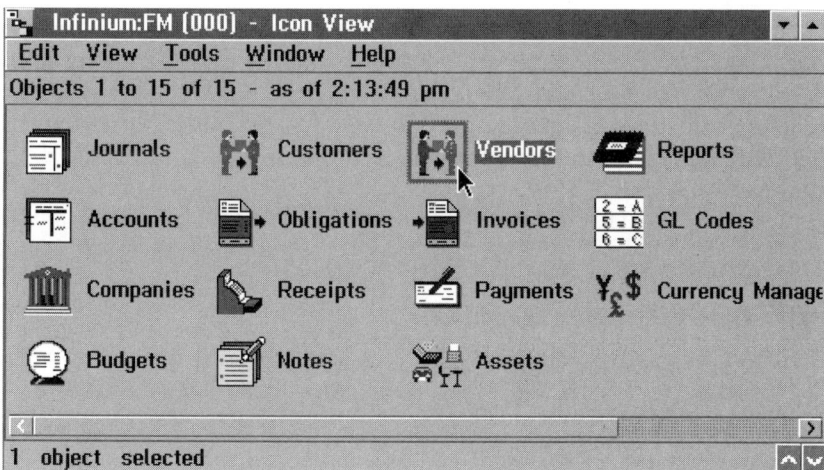

© Software 2000 Inc. 1996

Business object accounting systems depend on object request brokers (ORBs) for managing their message-based interobject communication. The battleground for control of this crucial component of business object applications is already split into two camps: a consortium of vendors supporting the Common Object Request Broker Architecture (CORBA) standards, and Microsoft Corporation, with its Distributed Component Object Model (DCOM). Although the few accounting vendors that have business-object-based applications support CORBA standards, Microsoft's early 1997 release of its Transaction Server (a DCOM ORB) is likely to gain the support of many accounting vendors, judging by initial reaction to the product.

ScreenHelp: This screen shows the main menu for Software 2000's Infinium Financials product. The interface is based on the use of standard business objects that represent the people, places, and things of the accounting processes the application manages. You work with a business object's functionality by clicking on the object icon to display object-specific data editors and viewers or by dragging one business object onto another to access specific behaviors based on the combination of objects selected.

Applets

Components allow accounting software to take full advantage of new computing paradigms such as mobile computing and Internet computing. When an accounting system's functions consist of small components, the software is easier to deploy on laptops, within Web pages, and by e-mail. Because components deliver only a subset of functionality, they require less memory and fewer disk resources than full-blown applications. Consequently, they download faster across communication infrastructures such as the Internet and are easier to deploy and use on limited-resource computers such as laptops.

In Internet accounting, components are often called applets. You can store applets on a Web server, download them to any Internet/intranet client, and embed them in Web documents accessed by Web browsers without implementing the entire resource-hungry functionality of an accounting module. Many leading accounting vendors have released applets for use in Internet browsers. In 1997 you can expect this trickle to become a torrent as vendors provide applets to handle

- submission of travel and expense timesheets
- budget entry, revision, and review
- order, requisition, and payment approvals
- balance, comparative, aging, and audit-trail queries
- dissemination of accounting reports and documents

- self-service vendor or customer account maintenance

- decision support and analysis of business statistics or key performance indicators

Making applets available to remote laptops or over the Internet will expand the accounting domain and extend the reach of accounting systems. As a result, accounting data will become more valuable, and far better use will be made of accounting information.

Component Accounting Using OLE

Frameworks, business objects, and applets are being developed to deliver the component accounting systems of the future, but applications that comply with Microsoft's Object Linking and Embedding (OLE) 2.0 specification deliver a form of component accounting today. OLE was conceived as a way to build compound documents consisting of several discrete functional objects, each of which can be edited in place. So, for example, a single document page might include text, a spreadsheet, and a chart. Each component may be created by a different application (such as Microsoft Word or Excel); the document page acts simply as a container for the objects. Double-clicking the mouse on an object lets you edit it right on the page, using the application that created it. Changing a value in one component object can force another to be refreshed. So changing a number in text, for example, may force a spreadsheet to be recalculated and a chart to be refreshed to reflect the changed value.

In its earliest implementations, OLE was essentially application collaboration technology. From those compound-document and application-integration roots, OLE has become the basis for Microsoft's distributed business object model, which means OLE compliance is becoming an important attribute for accounting systems to offer. OLE now supports visual and nonvisual objects and communication between them across networks. An example of a visual object might be a graphical display of a calendar you can flip through on screen to select a date to be inserted into a form, such as an invoice. The calendar object could be located on a client or server and could be used by many different accounting modules for various purposes.

A nonvisual object might be a balance calculator that works out the inception-, year-, period-, week-, and day-to-date balance for an account based on a date selected from the calendar object. The user would never see the calculator; only the results of its calculations. A nonvisual object can also reside on either a client or server machine located anywhere on the network. An OLE-compliant application maintains a registry of information about objects and their locations. It's easy to see how accounting systems can benefit from being designed around the paradigm of visual and nonvisual objects. State Of The Art Software, with its Acuity Financials product, is among the first to exploit this approach in the design of a product from the ground up.

Many accounting vendors use OLE to provide seamless integration between accounting components and third-party systems, such as

- passing data to and from third-party worksheets embedded in accounting transaction entry screens for use in budgeting or allocations

- passing data to and from third-party charting engines to provide graphs and charts that change dynamically based on the results of queries against the accounting data

- linking accounting transactions, such as invoices, or accounting records, such as inventory items, to documents or pictures managed by third-party imaging products

OLE has already extended the functional boundaries of accounting systems in dozens of ways and is likely to become even more influential in its new guise as ActiveX.

Conclusion

Software industry leaders, such as Microsoft and IBM, and leading accounting vendors, such as SAP, are working on component frameworks for specific businesses and line-of-business objects that are likely to have a significant impact on the accounting sector. Frameworks for use in financial, supply-chain, and manufacturing applications will probably be some of the first to appear due to their wide applicability. Initially the targets for these software components will be internal IS departments in large corporations that want to build their own in-house accounting systems using rapid application development techniques, or value-added resellers who will custom-ize the frameworks for use in specific niche businesses. But eventually, you can expect third-party vendors to incorporate these components as part of their own systems built specifically for resale.

A whole aftermarket is also likely to develop around these frameworks and objects, just as it already has around development tools and accounting applications. Small "boutique" programming companies will add value to frameworks and business objects by customizing them for niche businesses or by adding new methods or enhancing existing methods. Systems will no longer be priced by module, but by component, with logical collections of components being sold under traditional classifications, such as "GL module" or "AP module." Systems won't be upgraded in major releases, but incrementally, object by object, allowing for better functional scalability and a less costly upgrade process.

QUESTIONS FOR VENDORS

1. If the vendor uses the term "business object," what exactly does it mean? Does it mean platform-independent, free-standing objects encapsulating attributes and methods that communicate via messages or visual and nonvisual OLE or ActiveX components limited to Microsoft platforms? Or does the vendor mean some other, more nebulous concept?

2. Can the software be purchased only as a suite or as modules? Or can it be purchased in smaller functional components or applets, such as requisition entry or order inquiry?

3. Are software upgrades delivered for the whole application, module-by-module, or function-by-function?

4. If the software is based on a framework of business objects, can they be used to modify and enhance the core framework functions without modifying the application at the source code level?

5. Does the software use conventional menu-based functions, or is it based on business objects that represent people, places and things familiar to us in the real world?

6. What is the vendor's position on CORBA or DCOM compliance, and is that position reflected in the vendor's software or third-party partnerships?

Chapter 11

Segmenting the Market

Before you start to compile a long-list of client/server accounting vendors, it helps to be familiar with the four major segments of the accounting software market. By understanding which segment best describes your situation, you can avoid wasting time looking at products designed for businesses whose needs are entirely different from yours. The four classes of accounting solutions are

1. SOHO — the small office, home office segment, in which one person is usually responsible for all the business accounting.

2. Workgroup — businesses that may need additional accounting staff, such as receivables or payables clerks, because the business generates many more transactions than the SOHO segment.

3. Corporate — businesses that generally have separate workgroups of accounting users focused on specific business processes, such as procurement or fulfillment.

4. Enterprise — businesses that have many subsidiaries and thus have a broader range of functional needs and demand more sophisticated financial reporting.

Table 11.1 provides details about each of these four major market segments.

It can be argued that the corporate segment could be further segmented into self-standing or subsidiary corporates. A corporate that is a subsidiary of a larger operation has specialized functional needs, such as having to report data to its corporate parent, manage intercompany transactions, and

Table 11.1 Segmenting the Accounting Software Market

Characteristic	SOHO	Workgroup	Corporate	Enterprise
CUSTOMERS				
Annual revenue	< $1 million	$1 - $50 million	$50 - $500 million	$500 million +
Number of users	1	2 - 25	25 - 250	>250
Remote users?	No	No	Yes	Yes
Monthly transaction volume	< 500	500-5,000	5,000-50,000	>50,000
Sites per location	Single/domestic	Single/domestic	Multi/international	Multi/international
APPLICATIONS				
Price per module	< $1000 (per system)	$1,000 - $5,000	$5,000 - $50,000	> $50,000
Typical 5-module system cost	< $1000	< $25,000	< $250,000	> $250,000
Typical implementation cost per dollar of software cost	n/a	$1 - $2	$2 - $3	> $3
Typical implementation times	Days	Weeks	Months	Years
Distribution	Retail/mail order	Retail/VAR	VAR/Direct	Direct only
TECHNOLOGY				
Server operating platforms	MS-DOS/Mac OS	Novell NetWare/MS NT Server/SCO Unix	AS/400/MS NT Server/Unix	AS/400/ mainframe/Unix
Database	Btrieve/MS Jet/ISAM	Btrieve/MS Jet/MS SQL Server/Cobol ISAM	RDBMS (Oracle, Sybase,Informix, IBM DB2, MS SQL Server)/Cobol or C ISAM/IBM OS/400	RDBMS (Oracle, Sybase, Informix, IBM DB2)/Cobol or C ISAM/IBM OS/400
Database size	<1 MB	<100 MB	<1 GB	>1 GB
User interface	MS Windows/Mac OS	MS Windows/Mac OS	MS Windows/OSG Motif/Character	MS Windows/OSG Motif/Character
Implementation resource	Independent consultant	Independent consultant/VAR/vendor	Accounting firm/VAR/ vendor	Big 6 firm/systems integrator/vendor
FUNCTIONALITY				
Financials	•	•	•	•
Consolidation reporting			•	•
Distribution/supply chain		•	•	•
Manufacturing				•
Payroll	•	•	•	•
Human resources			•	•
Multicurrency			•	•
Multilingual			•	•
Allocations			•	•
Intercompany			•	•
Treasury management			•	•
EDI				•

receive "push-down" allocations for use of headquarters or regional office services.

The difficulty with segmenting client/server accounting software is that all packages, whether workgroup, corporate, or enterprise, share similar server platforms, RDBMS products, and GUIs. Consequently, it's hard to differentiate between products based on these criteria. In the past, because accounting software was much more platform-specific, your platform choice essentially determined your range of options. Now there are dozens of systems, for example, that support similar relational databases on a variety of Unix platforms with a Microsoft Windows front-end GUI. So what's the difference between packages?

With newly released client/server accounting packages, some new differentiators to be concerned with are

- When was the package first released?

- What is the breadth and depth of the package's functionality?

- How closely does the package support your key business processes?

- How well does the package leverage its client, server, database, and network technology platforms?

- How many useful third-party applications can be integrated with the application?

- Who are the vendor's main technology and marketing partners?

- Which de facto and de jure standards does the package support?

- How is the package being adapted to meet the opportunity of the Internet?

- Is the GUI design consistent with the GUI of your other desktop applications?

- How usable is the package on a day-to-day basis?

- How easy is it to move your data from your legacy system database into the database of the new application?

Many vendors previously positioned in the workgroup sector have released new client/server product suites that let them sell to larger corporate users. Similarly, many vendors previously positioned at the enterprise level are now making their packages available on less expensive operating platforms to make them more attractive buys for nonenterprise-level businesses. Table 11.2 lists some of the leading vendors servicing each of these segments.

Table 11.2. Some Leading Vendors in the Major Market Segments

Segment	Vendor	Product
SOHO	Intuit	Quicken, QuickBooks
	Microsoft	Money
	Peachtree	Peachtree Accounting
	Bestware	MYOB
Workgroup	Great Plains Software	Dynamics
	Platinum Software	Platinum for Windows
	Solomon Software	Solomon IV For Windows
	State Of The Art	MAS 90
Corporate	CODA	CODA Financials
	Lawson	Lawson Insight
	Software 2000	Infinium
	Systems Union	SunSystems
Enterprise	JD Edwards	OneWorld
	Oracle	Oracle Applications
	PeopleSoft	PeopleSoft 6
	SAP	R/3
	SSA	BPCS

Positioning a prospective vendor in terms of the market sector its products are designed to serve will help you make a cost-effective software decision. The cost differential between segments can be an order of magnitude or more, not just in terms of software acquisition, but also for the platform required to run the software and the consulting time needed to implement it.

Chapter 12

Twenty Technology Questions to Ask Vendors

Here are 20 questions you should ask vendors to answer as part of your technology "due diligence" process. These questions focus on the vendor's technology offering, as opposed to its functional offering, and are more general than the questions at the end of Chapters 1 through 10. The relevance of the questions below, of course, will depend entirely on your business and system needs, so you should regard them only as guidelines. But at least if you know the vendor's "story" on all these questions, you'll have a basic idea of which vendors have a better technology fit for your needs before you begin an in-depth functional evaluation. I've also suggested a simple three-point weighting system to give you a way to grade vendor responses.

1. **If client/server, can the application be deployed in a two-tier, three-tier, or n-tier configuration?**

Answer	Score
Two-tier (thin or thick client configurations)	1
Three-tier (middle tier of application servers)	2
N-tier (multiple middle tiers of application and process servers)	3

   If you don't see the value of a three-tier or n-tier architecture in your business, you may want to reverse the scores for this question.

2. **What primary methodology does the client/server application use to access its database?**

Answer	Score
Embedded SQL queries or calls to third-party ODBC drivers	1
Hand-tuned ODBC or native drivers	2
SQL RPC or ODBC calls to database stored procedures	3

If the vendor supports multiple databases, it may not support the use of stored procedures in all of them. But if the vendor supports only one database, it should make optimum use of the stored procedure functionality available for that database.

3. **How adaptable is the application to accessing data in different databases or combinations of databases?**

Answer	Score
Supports one database engine only	1
Supports more than one database engine	2
Supports linking to more than one database engine within an application	3

If you anticipate needing support for only one database engine, reverse the scores for this question.

4. **How adaptable is the application to supporting different user interfaces on the desktop?**

Answer	Score
Supports Win 95 and is certified with the Designed for Windows 95 logo	1
As above, plus supports Windows 3.1 16-bit client	2
Supports other user interfaces in addition to Microsoft Windows	3

If you require support only for Microsoft Windows 95 desktops, you can reverse the scores for this question.

5. **How adaptable is the application to running on different database, application, or process-server platforms?**

Answer	Score
Supports Microsoft NT Server platform	1
Supports Microsoft NT Server plus one or more variants of Unix	2
Supports NT, Unix, and other platforms (e.g., IBM AS/400 or Novell NetWare)	3

If you need support only for Microsoft NT Server servers, you can reverse the scores for this question.

6. How adaptable is the application to user customization?

Answer	Score
Written in a commercial 4GL with accessible code	1
Written in a compiled language, but screen forms, fields, and menus can be modified	2
As above, but includes fully integrated development environment (IDE)	3

If you want to purchase an application written in a particular 4GL used elsewhere in your organization, reverse the scores for this question.

7. Is the application e-mail-enabled?

Answer	Score
Links to a popular e-mail protocol such as Microsoft MAPI	1
Links to a popular e-mail protocol and to an Internet protocol such as SMTP	2
E-mail also hooks in to a built-in alert or workflow notification engine	3

8. Is the application image-enabled, and does it support document management?

Answer	Score
Allows images to be linked to reference data such as inventory or employees	1
As above, plus lets you attach document images to transactions	2
As above, plus allows document images to be part of routing workflows	3

9. **Is the application workflow-enabled?**

Answer	Score
Supports workflow processes within a conventional accounting module	1
Supports workflow processes across accounting modules	2
Supports workflow processes that cross third-party applications	3

If you need only limited single-module workflows, you should reverse the scores for this question.

10. **Is the application telephony-enabled?**

Answer	Score
Will autodial a telephone number displayed on a screen form	1
Provides touch-tone access to the accounting database for phone inquiries	2
Supports touch tone access to initiate or participate in workflows	3

11. **Is the application Internet-enabled?**

Answer	Score
Supports report publication in HTML formats and Internet e-mail distribution	1
Supports transaction initiation or queries using forms in HTML pages	2
As above, plus some form of business-to-business or consumer-to-business electronic commerce capability	3

12. **Is the application OLAP-enabled?**

Answer	Score
Supports a third-party OLAP application via a metadata or data dictionary layer	1
Supports automated data transfer to an OLAP engine such as Arbor Essbase	2
Includes a fully integrated OLAP module as a deliverable or option	3

If the application need integrate only with an OLAP application used elsewhere in your organization, you can reverse the scores for this question.

13. Does the application offer a broad and deep range of modules?

Answer	Score
Supports financials only	1
Supports financials and supply-chain modules	2
Supports financials, supply-chain modules, and manufacturing or human resources	3

If you need to implement financials only, you should reverse the scores for this question.

14. Is the application focused on Microsoft technology, and is the vendor closely aligned to Microsoft commercial standards?

Answer	Score
Certified with Designed for Windows 95 logo	1
Certified with Office Compatible logo	2
Certified with Designed for BackOffice logo	3

If you are not impressed by Microsoft logos, you can ignore this question.

15. How is the application sold and supported?

Answer	Score
Direct only	1
Through value-added resellers (VARs), Big Six accounting firms, or systems integrators	2
Direct and through VARs, Big Six accounting firms, or systems integrators	3

16. How deep is the application's multinational accounting functionality?

Answer	Score
Supports multicurrency accounting in one or more modules	1
As above, plus consolidation reporting and FAS52-compliant translations	2
As above, plus multilingual or dual-language versions	3

If you don't need multinational functionality, ignore this question.

17. Is the application available in multiple languages?

Answer	Score
U.S. English only	1
Multi-English versions (e.g., for UK, South Africa, Australia)	2
One or more non-English versions	3

If you don't require multilingual functionality, you can ignore this question.

18. How is the application supported and upgraded?

Answer	Score
Telephone, fax, and mail	1
As above, plus online, Internet, or Lotus Notes-based support forums	2
As above, plus by way of an integrated support and problem resolution module in the application	3

19. How many live sites are running the application?

Answer	Score
Fewer than 100	1
Fewer than 500	2
More than 500	3

20. What is the implementation cost of the application?

Answer	Score
More than $5 per dollar of software cost	1
From $2 to $5 per dollar of software cost	2
Less than or equal to the cost of the software	3

Chapter 13

Vendor
Profiles

Each of the 54 accounting-vendor profiles that follow provides a simple, one-page overview of one vendor's client/server accounting offering. The profiles include most of the leading accounting vendors and products included in Software Magazine's Top 100 list of independent software companies and are representative of the U.S. accounting software market in 1996.

The vendors themselves provided and checked the data for these profiles; I have not verified the data independently. The client/server accounting market changes quickly, so some information in the vendor profiles will undoubtedly become incorrect over time. Don't make a purchasing or even a short-listing decision based only on the information in these profiles; always verify your key criteria with vendors before you include them or remove them from your evaluation shortlist.

Each vendor profile is divided into five main sections: vendor data, product suite, client/server platform, technology support, and marketing information.

Vendor Data

This section lists the vendor's address and contact information plus some basic facts about the vendor and its fiscal 1995 or 1996 revenues, if disclosed. Telephone and fax numbers can change, so you may find it worthwhile to try to verify them on the vendor's Web site before using them.

Product Suite

This section shows the name of the product and the range of modules in the product suite, divided into Financial, Supply Chain, and Other categories. Note that some of the modules, such as payroll, fixed assets, or EDI software, may be supplied by third parties rather than by the primary vendor. The cross-vendor lists provided in Chapter 14 let you easily compare the offerings of each vendor by category.

Client/Server Platform Support

This section shows the range of server, database, and GUI platforms the vendor supports. The supported platforms can change with every new release. Usually options are added, but sometimes vendors drop a platform option that is no longer commercially viable. Chapter 14 includes a cross-vendor list for comparing platform support.

Technology Support

This section shows what technology was used to develop the product, what additional application technology the product supports, and whether it supports certain Microsoft technologies and initiatives. The development technology used can have a big influence on the product's performance, how easy it is to customize, and how quickly it can be enhanced. Chapter 14 includes a cross-vendor checklist for comparing development, application, and Microsoft-technology support.

Marketing Information

This section shows some pricing, distribution-channel, and international support data for the product. The module pricing listed in the vendor profile is a very rough guideline only — individual modules are often priced differently from each other or cross the pricing boundaries used in the checklists. Also, pricing is frequently negotiable, which means that starting prices are merely indicative of the package's relative pricing position. Again, in Chapter 14 you'll find a cross-vendor list to help you compare the products based on this marketing information.

Checklist Legend

■ = Yes, supported in shipping software

☐ = No, not supported in shipping software

▤ = Third-party software

AccountMate Software

Address Data
AccountMate Software
20 Sunnyside Drive
Mill Valley CA 94941

Contact Data
Tel: (415) 381-1011
Fax: (415) 381-6902
Web: http://www.accountmate.com

Facts and Figures
Founded: 1984
Country: USA
Status: Private
WW User Base: 85,000+
'95 Revenues ($M): N/A

Visual AccountMate C/S

Financials
- ■ General Ledger
- ■ Financial Reporting
- ■ Consolidations
- ■ Budgeting
- □ Project Management
- □ Fixed Assets

Supply Chain
- ■ Accounts Payable
- ■ Purchasing
- ■ Accounts Receivable
- ■ Order Entry   ■ Billing
- ■ Inventory   ■ Bill of Materials
- □ EDI

Other
- ■ Manufacturing
- □ Human Resources
- ■ Payroll & Benefits
- ■ Multi Currency
- ■ Multi Lingual
- ■ Allocations

C/S PLATFORM SUPPORT

Server Operating System
- □ UNIX - IBM
- □ UNIX - HP
- □ UNIX - SUN
- ■ Microsoft NT Server
- ■ Novell Netware
- ■ IBM AS/400 (OS/400)
- □ IBM Mainframe

RDBMS Engine
- □ Oracle 7
- □ Informix
- □ Sybase 11
- ■ Microsoft SQL Server 6.x
- □ Centura SQLBase
- □ BTI Scaleable SQL/Btrieve
- ■ IBM DB2 (any platform)

Client GUI
- ■ Windows 95/NT (32 bit)
- □ Windows 3.x (16 bit)
- □ Apple MACOS
- □ IBM OS/2 Warp (32 bit)
- □ OSF/Motif
- □ Sun OpenLook
- □ Character

TECHNOLOGY SUPPORT

Development Technology
- □ C   □ C++
- □ COBOL   □ RPG
- □ SmallTalk □ Other OO
- □ Progress   □ NewEra
- □ PowerBuilder □ SQL Windows
- □ MS Access □ Visual Basic
- ■ OtherMS Visual FoxPro

Application Technology
- ■ E mail enabled
- ■ Workflow enabled
- □ Document/image enabled
- □ Telephony enabled
- □ OLAP connectivity or module
- □ User accessible Data Dictionary
- ■ Internet/Intranet access enabled

Microsoft Technology
- ■ MAPI Compliant
- □ MFC Used
- ■ ODBC Compliant
- ■ OLE 2/ActiveX Complaint
- ■ Designed for Windows 95 Logo
- ■ Office Compatible Logo
- ■ BackOffice Powered Logo

MARKETING INFORMATION

Start Price Per Module
- ■ < US$25K
- □ < US$50K
- □ < US$100K
- □ > US$100K

Sales Channel
- □ Direct Only
- ■ VAR Only
- □ Direct and VAR
- □ Big Six / Outsource Firm

Direct Sales & Support Offices
- ■ In USA/Canada
- □ In UK/Europe
- ■ In Pacific Rim/Australasia
- ■ In Central/South America

Agresso Corp.

Address Data

Agresso Corp.
1250 Oakmead Parkway
Sunnyvale, CA 94088

Contact Data

Tel: (408) 524-2910
Fax: (408) 524-2917
Web: http://www.agresso.com

Facts and Figures

Founded: 1991
Country: Norway
Status:
WW User Base: 500+
'95 Revenues ($M): N/A

Agresso Financials

Financials
- ■ General Ledger
- ■ Financial Reporting
- ■ Consolidations
- ■ Budgeting
- ■ Project Management
- ■ Fixed Assets

Supply Chain
- ■ Accounts Payable
- ■ Purchasing
- ■ Accounts Receivable
- ■ Order Entry
- ■ Inventory
- ■ EDI
- ■ Billing
- □ Bill of Materials

Other
- □ Manufacturing
- ■ Human Resources
- ■ Payroll & Benefits
- ■ Multi Currency
- ■ Multi Lingual
- ■ Allocations

C/S PLATFORM SUPPORT

Server Operating System
- ■ UNIX - IBM
- ■ UNIX - HP
- ■ UNIX - SUN
- ■ Microsoft NT Server
- ■ Novell Netware
- □ IBM AS/400 (OS/400)
- □ IBM Mainframe

RDBMS Engine
- ■ Oracle 7
- ■ Informix
- ■ Sybase 11
- ■ Microsoft SQL Server 6.x
- □ Centura SQLBase
- □ BTI Scaleable SQL/Btrieve
- □ IBM DB2 (any platform)

Client GUI
- ■ Windows 95/NT (32 bit)
- ■ Windows 3.x (16 bit)
- □ Apple MACOS
- □ IBM OS/2 Warp (32 bit)
- □ OSF/Motif
- □ Sun OpenLook
- ■ Character

TECHNOLOGY SUPPORT

Development Technology
- ■ C
- □ COBOL
- □ SmallTalk
- □ Progress
- □ PowerBuilder
- □ MS Access
- □ Other 4GL
- ■ C++
- □ RPG
- □ Other OO
- □ NewEra
- □ SQL Windows
- ■ Visual Basic

Application Technology
- ■ E mail enabled
- □ Workflow enabled
- ■ Document/image enabled
- □ Telephony enabled
- □ OLAP connectivity or module
- ■ User accessible Data Dictionary
- ■ Internet/Intranet access enabled

Microsoft Technology
- ■ MAPI Compliant
- ■ MFC Used
- □ ODBC Compliant
- □ OLE 2/ActiveX Complaint
- □ Designed for Windows 95 Logo
- □ Office Compatible Logo
- □ BackOffice Powered Logo

MARKETING INFORMATION

Start Price Per Module
- ■ < US$25K
- □ < US$50K
- □ < US$100K
- □ > US$100K

Sales Channel
- □ Direct Only
- □ VAR Only
- ■ Direct and VAR
- □ Big Six / Outsource Firm

Direct Sales & Support Offices
- ■ In USA/Canada
- ■ In UK/Europe
- □ In Pacific Rim/Australasia
- □ In Central/South America

American Software

Address Data
American Software
470 East Paces Ferry Road
Atlanta GA 30305

Contact Data
Tel: (404) 264-5296
Fax: (404) 264-5206
Web: http://www.amsoftware.com

Facts and Figures
Founded: 1970
Country: USA
Status: Public
WW User Base: 1600
'95 Revenues ($M): 79,462

Supply Chain Management

Financials
- ■ General Ledger
- ■ Financial Reporting
- ■ Consolidations
- ■ Budgeting
- ■ Project Management
- □ Fixed Assets

Supply Chain
- ■ Accounts Payable
- ■ Purchasing
- ■ Accounts Receivable
- ■ Order Entry ■ Billing
- ■ Inventory   ■ Bill of Materials
- ■ EDI

Other
- ■ Manufacturing
- □ Human Resources
- □ Payroll & Benefits
- ■ Multi Currency
- ■ Multi Lingual
- ■ Allocations

C/S PLATFORM SUPPORT

Server Operating System
- ■ UNIX - IBM
- ■ UNIX - HP
- ■ UNIX - SUN
- ■ Microsoft NT Server
- ■ Novell Netware
- ■ IBM AS/400 (OS/400)
- ■ IBM Mainframe

RDBMS Engine
- ■ Oracle 7
- □ Informix
- □ Sybase 11
- □ Microsoft SQL Server 6.x
- □ Centura SQLBase
- □ BTI Scaleable SQL/Btrieve
- ■ IBM DB2 (any platform)

Client GUI
- ■ Windows 95/NT (32 bit)
- ■ Windows 3.x (16 bit)
- □ Apple MACOS
- ■ IBM OS/2 Warp (32 bit)
- □ OSF/Motif
- □ Sun OpenLook
- □ Character

TECHNOLOGY SUPPORT

Development Technology
- □ C   ■ C++
- ■ COBOL   □ RPG
- □ SmallTalk   □ Other OO
- □ Progress   □ NewEra
- □ PowerBuilder □ SQL Windows
- □ MS Access   ■ Visual Basic
- □ Other 4GL

Application Technology
- □ E mail enabled
- ■ Workflow enabled
- ■ Document/image enabled
- ■ Telephony enabled
- □ OLAP connectivity or module
- ■ User accessible Data Dictionary
- □ Internet/Intranet access enabled

Microsoft Technology
- □ MAPI Compliant
- □ MFC Used
- □ ODBC Compliant
- □ OLE 2/ActiveX Compliant
- □ Designed for Windows 95 Logo
- □ Office Compatible Logo
- □ BackOffice Powered Logo

MARKETING INFORMATION

Start Price Per Module
- □ < US$25K
- ■ < US$50K
- □ < US$100K
- □ > US$100K

Sales Channel
- ■ Direct Only
- □ VAR Only
- □ Direct and VAR
- ■ Big Six / Outsource Firm

Direct Sales & Support Offices
- ■ In USA/Canada
- ■ In UK/Europe
- ■ In Pacific Rim/Australasia
- ■ In Central/South America

Apprise Software Inc.

Address Data
Apprise Software Inc.
3121 Route 22
Somerville NJ 08876

Contact Data
Tel: (908) 725-6000
Fax: (908) 725-9555
Web: http://www.apprisesi.com

Facts and Figures
Founded: 1984
Country: USA
Status: Private
WW User Base: N/A
'95 Revenues ($M): N/A

Apprise Financials

Financials
- ■ General Ledger
- ■ Financial Reporting
- ■ Consolidations
- ■ Budgeting
- ■ Project Management
- ■ Fixed Assets

Supply Chain
- ■ Accounts Payable
- ■ Purchasing
- ■ Accounts Receivable
- ■ Order Entry   ■ Billing
- ■ Inventory   □ Bill of Materials
- ■ EDI

Other
- □ Manufacturing
- □ Human Resources
- □ Payroll & Benefits
- ■ Multi Currency
- ■ Multi Lingual
- ■ Allocations

C/S PLATFORM SUPPORT

Server Operating System
- ■ UNIX - IBM
- ■ UNIX - HP
- ■ UNIX - SUN
- ■ Microsoft NT Server
- ■ Novell Netware
- ■ IBM AS/400 (OS/400)
- □ IBM Mainframe

RDBMS Engine
- ■ Oracle 7
- □ Informix
- □ Sybase 11
- □ Microsoft SQL Server 6.x
- □ Centura SQLBase
- □ BTI Scaleable SQL/Btrieve
- ■ IBM DB2 (any platform)

Client GUI
- ■ Windows 95/NT (32 bit)
- ■ Windows 3.x (16 bit)
- □ Apple MACOS
- □ IBM OS/2 Warp (32 bit)
- □ OSF/Motif
- □ Sun OpenLook
- □ Character

TECHNOLOGY SUPPORT

Development Technology
- □ C
- □ COBOL
- □ SmallTalk
- ■ Progress
- □ PowerBuilder
- □ MS Access
- □ Other 4GL
- □ C++
- □ RPG
- □ Other OO
- □ NewEra
- □ SQL Windows
- □ Visual Basic

Application Technology
- ■ E mail enabled
- ■ Workflow enabled
- ■ Document/image enabled
- □ Telephony enabled
- □ OLAP connectivity or module
- ■ User accessible Data Dictionary
- □ Internet/Intranet access enabled

Microsoft Technology
- ■ MAPI Compliant
- □ MFC Used
- ■ ODBC Compliant
- □ OLE 2/ActiveX Complaint
- □ Designed for Windows 95 Logo
- □ Office Compatible Logo
- □ BackOffice Powered Logo

MARKETING INFORMATION

Start Price Per Module
- ■ < US$25K
- □ < US$50K
- □ < US$100K
- □ > US$100K

Sales Channel
- □ Direct Only
- □ VAR Only
- □ Direct and VAR
- □ Big Six / Outsource Firm

Direct Sales & Support Offices
- ■ In USA/Canada
- ■ In UK/Europe
- □ In Pacific Rim/Australasia
- □ In Central/South America

Baan Company

Address Data
Baan Company
4600 Bohannon Ave.
Menlo Park CA 94025

Contact Data
Tel: (415) 462-4949
Fax: (415) 462-4953
Web: http://www.baan.com

Facts and Figures
Founded: 1978
Country: Netherlands
Status: Public
WW User Base: 1900
'95 Revenues ($M): 216.2

BAAN IV

Financials
- ■ General Ledger
- ■ Financial Reporting
- ■ Consolidations
- ■ Budgeting
- ■ Project Management
- ■ Fixed Assets

Supply Chain
- ■ Accounts Payable
- ■ Purchasing
- ■ Accounts Receivable
- ■ Order Entry ■ Billing
- ■ Inventory ■ Bill of Materials
- ■ EDI

Other
- ■ Manufacturing
- ☐ Human Resources
- ☐ Payroll & Benefits
- ■ Multi Currency
- ■ Multi Lingual
- ☐ Allocations

C/S PLATFORM SUPPORT

Server Operating System
- ■ UNIX - IBM
- ■ UNIX - HP
- ■ UNIX - SUN
- ■ Microsoft NT Server
- ■ Novell Netware
- ☐ IBM AS/400 (OS/400)
- ☐ IBM Mainframe

RDBMS Engine
- ■ Oracle 7
- ■ Informix
- ■ Sybase 11
- ■ Microsoft SQL Server 6.x
- ☐ Centura SQLBase
- ☐ BTI Scaleable SQL/Btrieve
- ☐ IBM DB2 (any platform)

Client GUI
- ■ Windows 95/NT (32 bit)
- ☐ Windows 3.x (16 bit)
- ■ Apple MACOS
- ☐ IBM OS/2 Warp (32 bit)
- ■ OSF/Motif
- ☐ Sun OpenLook
- ☐ Character

TECHNOLOGY SUPPORT

Development Technology
- ☐ C
- ☐ COBOL
- ☐ SmallTalk
- ☐ Progress
- ☐ PowerBuilder
- ☐ MS Access
- ■ Other 4GL: Triton
- ☐ C++
- ☐ RPG
- ☐ Other OO
- ☐ NewEra
- ☐ SQL Windows
- ☐ Visual Basic

Application Technology
- ☐ E mail enabled
- ☐ Workflow enabled
- ☐ Document/image enabled
- ☐ Telephony enabled
- ■ OLAP connectivity or module
- ■ User accessible Data Dictionary
- ■ Internet/Intranet access enabled

Microsoft Technology
- ☐ MAPI Compliant
- ☐ MFC Used
- ☐ ODBC Compliant
- ☐ OLE 2/ActiveX Complaint
- ☐ Designed for Windows 95 Logo
- ☐ Office Compatible Logo
- ■ BackOffice Powered Logo

MARKETING INFORMATION

Start Price Per Module
- ☐ < US$25K
- ☐ < US$50K
- ☐ < US$100K
- ■ > US$100K

Sales Channel
- ☐ Direct Only
- ☐ VAR Only
- ■ Direct and VAR
- ■ Big Six / Outsource Firm

Direct Sales & Support Offices
- ■ In USA/Canada
- ■ In UK/Europe
- ■ In Pacific Rim/Australasia
- ■ In Central/South America

BHR Software

Address Data
BHR Software
31 N. 2nd Street
San Jose CA 95113

Contact Data
Tel: (408) 882-0680
Fax: (408) 882-0690
Web: http://www.bhrsoftware.com

Facts and Figures
Founded: 1992
Country: USA
Status:
WW User Base: 27
'95 Revenues ($M): N/A

Financials
- ■ General Ledger
- ■ Financial Reporting
- ■ Consolidations
- ■ Budgeting
- ■ Project Management
- ☐ Fixed Assets

Supply Chain
- ■ Accounts Payable
- ■ Purchasing
- ■ Accounts Receivable
- ■ Order Entry ■ Billing
- ■ Inventory ■ Bill of Materials
- ■ EDI

Other
- ■ Manufacturing
- ☐ Human Resources
- ☐ Payroll & Benefits
- ☐ Multi Currency
- ☐ Multi Lingual
- ☐ Allocations

C/S PLATFORM SUPPORT

Server Operating System
- ■ UNIX - IBM
- ■ UNIX - HP
- ■ UNIX - SUN
- ■ Microsoft NT Server
- ☐ Novell Netware
- ☐ IBM AS/400 (OS/400)
- ☐ IBM Mainframe

RDBMS Engine
- ☐ Oracle 7
- ■ Informix
- ☐ Sybase 11
- ■ Microsoft SQL Server 6.x
- ☐ Centura SQLBase
- ☐ BTI Scaleable SQL/Btrieve
- ☐ IBM DB2 (any platform)

Client GUI
- ■ Windows 95/NT (32 bit)
- ☐ Windows 3.x (16 bit)
- ☐ Apple MACOS
- ☐ IBM OS/2 Warp (32 bit)
- ☐ OSF/Motif
- ☐ Sun OpenLook
- ■ Character

TECHNOLOGY SUPPORT

Development Technology
- ■ C
- ☐ COBOL
- ☐ SmallTalk
- ☐ Progress
- ☐ PowerBuilder
- ☐ MS Access
- ■ Other 4GL: Java
- ☐ C++
- ☐ RPG
- ☐ Other OO
- ☐ NewEra
- ☐ SQL Windows
- ☐ Visual Basic

Application Technology
- ■ E mail enabled
- ■ Workflow enabled
- ■ Document/image enabled
- ☐ Telephony enabled
- ☐ OLAP connectivity or module
- ☐ User accessible Data Dictionary
- ■ Internet/Intranet access enabled

Microsoft Technology
- ☐ MAPI Compliant
- ☐ MFC Used
- ■ ODBC Compliant
- ☐ OLE 2/ActiveX Complaint
- ☐ Designed for Windows 95 Logo
- ☐ Office Compatible Logo
- ■ BackOffice Powered Logo

MARKETING INFORMATION

Start Price Per Module
- ■ < US$25K
- ☐ < US$50K
- ☐ < US$100K
- ☐ > US$100K

Sales Channel
- ☐ Direct Only
- ☐ VAR Only
- ■ Direct and VAR
- ☐ Big Six / Outsource Firm

Direct Sales & Support Offices
- ■ In USA/Canada
- ☐ In UK/Europe
- ☐ In Pacific Rim/Australasia
- ☐ In Central/South America

Big Software

Address Data	**Contact Data**	**Facts and Figures**
Big Software	Tel: (408) 725-7200	Founded: 1993
1601 S. DeAnza Blvd.	Fax: (408) 725-7205	Country: USA
Cupertino CA 95014	Web: http://www.bigsoftware.com	Status: Private
		WW User Base: N/A
		'95 Revenues ($M): N/A

Big Business Server

Financials
- ■ General Ledger
- ■ Financial Reporting
- ☐ Consolidations
- ■ Budgeting
- ☐ Project Management
- ☐ Fixed Assets

Supply Chain
- ■ Accounts Payable
- ■ Purchasing
- ■ Accounts Receivable
- ■ Order Entry ■ Billing
- ■ Inventory ■ Bill of Materials
- ☐ EDI

Other
- ■ Manufacturing
- ☐ Human Resources
- ▣ Payroll & Benefits
- ☐ Multi Currency
- ☐ Multi Lingual
- ☐ Allocations

C/S PLATFORM SUPPORT

Server Operating System
- ☐ UNIX - IBM
- ☐ UNIX - HP
- ☐ UNIX - SUN
- ■ Microsoft NT Server
- ☐ Novell Netware
- ☐ IBM AS/400 (OS/400)
- ☐ IBM Mainframe

RDBMS Engine
- ☐ Oracle 7
- ☐ Informix
- ☐ Sybase 11
- ☐ Microsoft SQL Server 6.x
- ☐ Centura SQLBase
- ☐ BTI Scaleable SQL/Btrieve
- ☐ IBM DB2 (any platform)

Client GUI
- ■ Windows 95/NT (32 bit)
- ☐ Windows 3.x (16 bit)
- ■ Apple MACOS
- ☐ IBM OS/2 Warp (32 bit)
- ☐ OSF/Motif
- ☐ Sun OpenLook
- ☐ Character

TECHNOLOGY SUPPORT

Development Technology
- ■ C ■ C++
- ☐ COBOL ☐ RPG
- ☐ SmallTalk ☐ Other OO
- ☐ Progress ☐ NewEra
- ☐ PowerBuilder ☐ SQL Windows
- ☐ MS Access ☐ Visual Basic
- ☐ Other: 4D Server

Application Technology
- ■ E mail enabled
- ☐ Workflow enabled
- ☐ Document/image enabled
- ☐ Telephony enabled
- ☐ OLAP connectivity or module
- ☐ User accessible Data Dictionary
- ■ Internet/Intranet access enabled

Microsoft Technology
- ☐ MAPI Compliant
- ☐ MFC Used
- ☐ ODBC Compliant
- ☐ OLE 2/ActiveX Complaint
- ■ Designed for Windows 95 Logo
- ☐ Office Compatible Logo
- ☐ BackOffice Powered Logo

MARKETING INFORMATION

Start Price Per Module
- ■ < US$25K
- ☐ < US$50K
- ☐ < US$100K
- ☐ > US$100K

Sales Channel
- ☐ Direct Only
- ☐ VAR Only
- ■ Direct and VAR
- ☐ Big Six / Outsource Firm

Direct Sales & Support Offices
- ■ In USA/Canada
- ■ In UK/Europe
- ■ In Pacific Rim/Australasia
- ☐ In Central/South America

Carillon Financials Corporation

Address Data	**Contact Data**	**Facts and Figures**
Carillon Financials	Tel: (800) 739-9933	Founded: 1990
12750 Merit Dr., #1440	Fax: (972) 458-2977	Country: USA
Dallas TX 75251	Web: http://www.pettitco.com	Status:
		WW User Base: N/A
		'95 Revenues ($M): N/A

Carillon Financials

Financials
- ■ General Ledger
- ■ Financial Reporting
- ☐ Consolidations
- ■ Budgeting
- ☐ Project Management
- ■ Fixed Assets

Supply Chain
- ■ Accounts Payable
- ■ Purchasing
- ■ Accounts Receivable
- ■ Order Entry   ■ Billing
- ■ Inventory   ☐ Bill of Materials
- ■ EDI

Other
- ☐ Manufacturing
- ☐ Human Resources
- ☐ Payroll & Benefits
- ■ Multi Currency
- ■ Multi Lingual
- ☐ Allocations

C/S PLATFORM SUPPORT

Server Operating System
- ☐ UNIX - IBM
- ☐ UNIX - HP
- ☐ UNIX - SUN
- ■ Microsoft NT Server
- ■ Novell Netware
- ☐ IBM AS/400 (OS/400)
- ☐ IBM Mainframe

RDBMS Engine
- ■ Oracle 7
- ☐ Informix
- ■ Sybase 11
- ■ Microsoft SQL Server 6.x
- ☐ Centura SQLBase
- ☐ BTI Scaleable SQL/Btrieve
- ☐ IBM DB2 (any platform)

Client GUI
- ■ Windows 95/NT (32 bit)
- ■ Windows 3.x (16 bit)
- ☐ Apple MACOS
- ☐ IBM OS/2 Warp (32 bit)
- ☐ OSF/Motif
- ☐ Sun OpenLook
- ☐ Character

TECHNOLOGY SUPPORT

Development Technology
- ☐ C
- ☐ COBOL
- ☐ SmallTalk
- ☐ Progress
- ■ PowerBuilder
- ☐ MS Access
- ☐ Other: 4D Server
- ☐ C++
- ☐ RPG
- ☐ Other OO
- ☐ NewEra
- ☐ SQL Windows
- ☐ Visual Basic

Application Technology
- ☐ E mail enabled
- ☐ Workflow enabled
- ☐ Document/image enabled
- ☐ Telephony enabled
- ☐ OLAP connectivity or module
- ■ User accessible Data Dictionary
- ☐ Internet/Intranet access enabled

Microsoft Technology
- ■ MAPI Compliant
- ☐ MFC Used
- ■ ODBC Compliant
- ■ OLE 2/ActiveX Complaint
- ☐ Designed for Windows 95 Logo
- ☐ Office Compatible Logo
- ☐ BackOffice Powered Logo

MARKETING INFORMATION

Start Price Per Module
- ■ < US$25K
- ☐ < US$50K
- ☐ < US$100K
- ☐ > US$100K

Sales Channel
- ☐ Direct Only
- ☐ VAR Only
- ■ Direct and VAR
- ☐ Big Six / Outsource Firm

Direct Sales & Support Offices
- ■ In USA/Canada
- ☐ In UK/Europe
- ☐ In Pacific Rim/Australasia
- ☐ In Central/South America

CODA Inc.

Address Data	Contact Data	Facts and Figures
Address Data	**Contact Data**	**Facts and Figures**
CODA Inc.	Tel: (603) 647-9600	Founded: 1979
1155 Elm Street	Fax: (603) 647-2634	Country: England
Manchester NH 03101	Web: http://www.coda-financials.com	Status: Public
		WW User Base: 1400+
		'95 Revenues ($M): $52

CODA Financials

Financials
- ■ General Ledger
- ■ Financial Reporting
- ■ Consolidations
- ■ Budgeting
- ■ Project Management
- ■ Fixed Assets

Supply Chain
- ■ Accounts Payable
- ■ Purchasing
- ■ Accounts Receivable
- □ Order Entry □ Billing
- □ Inventory □ Bill of Materials
- □ EDI

Other
- □ Manufacturing
- □ Human Resources
- □ Payroll & Benefits
- ■ Multi Currency
- ■ Multi Lingual
- ■ Allocations

C/S PLATFORM SUPPORT

Server Operating System
- ■ UNIX - IBM
- ■ UNIX - HP
- ■ UNIX - SUN
- ■ Microsoft NT Server
- □ Novell Netware
- □ IBM AS/400 (OS/400)
- □ IBM Mainframe

RDBMS Engine
- ■ Oracle 7
- ■ Informix
- ■ Sybase 11
- ■ Microsoft SQL Server 6.x
- □ Centura SQLBase
- □ BTI Scaleable SQL/Btrieve
- □ IBM DB2 (any platform)

Client GUI
- ■ Windows 95/NT (32 bit)
- ■ Windows 3.x (16 bit)
- □ Apple MACOS
- □ IBM OS/2 Warp (32 bit)
- □ OSF/Motif
- □ Sun OpenLook
- □ Character

TECHNOLOGY SUPPORT

Development Technology
- ■ C   ■ C++
- □ COBOL   □ RPG
- □ SmallTalk   □ Other OO
- □ Progress   □ NewEra
- □ PowerBuilder   □ SQL Windows
- □ MS Access   ■ Visual Basic
- □ Other: 4GL

Application Technology
- ■ E mail enabled
- ■ Workflow enabled
- ■ Document/image enabled
- □ Telephony enabled
- ■ OLAP connectivity or module
- ■ User accessible Data Dictionary
- ■ Internet/Intranet access enabled

Microsoft Technology
- □ MAPI Compliant
- ■ MFC Used
- ■ ODBC Compliant
- ■ OLE 2/ActiveX Complaint
- □ Designed for Windows 95 Logo
- □ Office Compatible Logo
- □ BackOffice Powered Logo

MARKETING INFORMATION

Start Price Per Module
- □ < US$25K
- □ < US$50K
- □ < US$100K
- ■ > US$100K

Sales Channel
- □ Direct Only
- □ VAR Only
- ■ Direct and VAR
- □ Big Six / Outsource Firm

Direct Sales & Support Offices
- ■ In USA/Canada
- ■ In UK/Europe
- ■ In Pacific Rim/Australasia
- ■ In Central/South America

Computer Associates International Inc.

Address Data	**Contact Data**	**Facts and Figures**
Computer Associates Intl. Inc.	Tel: (516) 342-2245	Founded: 1976
One Computer Associates Plaza	Fax: (516) 342-5734	Country: USA
Islandia NY 11788-7000	Web: http://www.cai.com	Status: Public
		WW User Base: N/A
		'95 Revenues ($M): 2.6B

CA-Masterpiece

Financials
- ■ General Ledger
- ■ Financial Reporting
- ■ Consolidations
- ■ Budgeting
- ■ Project Management
- ■ Fixed Assets

Supply Chain
- ■ Accounts Payable
- ■ Purchasing
- ■ Accounts Receivable
- ☐ Order Entry   ☐ Billing
- ■ Inventory   ☐ Bill of Materials
- ■ EDI

Other
- ☐ Manufacturing
- ☐ Human Resources
- ☐ Payroll & Benefits
- ■ Multi Currency
- ■ Multi Lingual
- ■ Allocations

C/S PLATFORM SUPPORT

Server Operating System
- ■ UNIX - IBM
- ■ UNIX - HP
- ■ UNIX - SUN
- ■ Microsoft NT Server
- ☐ Novell Netware
- ■ IBM AS/400 (OS/400)
- ■ IBM Mainframe

RDBMS Engine
- ■ Oracle 7
- ■ Informix
- ☐ Sybase 11
- ☐ Microsoft SQL Server 6.x
- ☐ Centura SQLBase
- ☐ BTI Scaleable SQL/Btrieve
- ■ IBM DB2 (any platform)

Client GUI
- ■ Windows 95/NT (32 bit)
- ■ Windows 3.x (16 bit)
- ☐ Apple MACOS
- ☐ IBM OS/2 Warp (32 bit)
- ☐ OSF/Motif
- ☐ Sun OpenLook
- ■ Character

TECHNOLOGY SUPPORT

Development Technology
- ■ C
- ■ COBOL
- ☐ SmallTalk
- ☐ Progress
- ☐ PowerBuilder
- ☐ MS Access
- ☐ Other 4GL
- ■ C++
- ☐ RPG
- ■ Other OO
- ☐ NewEra
- ☐ SQL Windows
- ☐ Visual Basic

Application Technology
- ■ E mail enabled
- ■ Workflow enabled
- ☐ Document/image enabled
- ☐ Telephony enabled
- ☐ OLAP connectivity or module
- ☐ User accessible Data Dictionary
- ■ Internet/Intranet access enabled

Microsoft Technology
- ■ MAPI Compliant
- ■ MFC Used
- ☐ ODBC Compliant
- ☐ OLE 2/ActiveX Complaint
- ☐ Designed for Windows 95 Logo
- ☐ Office Compatible Logo
- ☐ BackOffice Powered Logo

MARKETING INFORMATION

Start Price Per Module
- ☐ < US$25K
- ☐ < US$50K
- ☐ < US$100K
- ☐ > US$100K

Sales Channel
- ☐ Direct Only
- ☐ VAR Only
- ■ Direct and VAR
- ☐ Big Six / Outsource Firm

Direct Sales & Support Offices
- ■ In USA/Canada
- ■ In UK/Europe
- ■ In Pacific Rim/Australasia
- ■ In Central/South America

Computron Software Inc.

Address Data
Computron Software Inc.
301 Route 17 North
Rutherford NJ 07070

Contact Data
Tel: (201) 935-3400
Fax: (201) 935-7678
Web: http://www.ctronsoft.com

Facts and Figures
Founded: 1978
Country: USA
Status: Public
WW User Base: 700
'93 Revenues ($M): 55.5

Computron Financials

Financials
- ■ General Ledger
- ■ Financial Reporting
- □ Consolidations
- □ Budgeting
- □ Project Management
- ■ Fixed Assets

Supply Chain
- ■ Accounts Payable
- ■ Purchasing
- ■ Accounts Receivable
- □ Order Entry   □ Billing
- ■ Inventory   □ Bill of Materials
- □ EDI

Other
- □ Manufacturing
- □ Human Resources
- □ Payroll & Benefits
- ■ Multi Currency
- ■ Multi Lingual
- □ Allocations

C/S PLATFORM SUPPORT

Server Operating System
- ■ UNIX - IBM
- ■ UNIX - HP
- ■ UNIX - SUN
- ■ Microsoft NT Server
- ■ Novell Netware
- □ IBM AS/400 (OS/400)
- □ IBM Mainframe

RDBMS Engine
- ■ Oracle 7
- ■ Informix
- ■ Sybase 11
- ■ Microsoft SQL Server 6.x
- □ Centura SQLBase
- □ BTI Scaleable SQL/Btrieve
- □ IBM DB2 (any platform)

Client GUI
- ■ Windows 95/NT (32 bit)
- ■ Windows 3.x (16 bit)
- □ Apple MACOS
- □ IBM OS/2 Warp (32 bit)
- □ OSF/Motif
- □ Sun OpenLook
- ■ Character

TECHNOLOGY SUPPORT

Development Technology
- ■ C   ■ C++
- ■ COBOL   □ RPG
- □ SmallTalk   □ Other OO
- □ Progress   □ NewEra
- □ PowerBuilder   □ SQL Windows
- □ MS Access   ■ Visual Basic
- ■ Other 4GL: NeuronData

Application Technology
- ■ E mail enabled
- ■ Workflow enabled
- ■ Document/image enabled
- □ Telephony enabled
- □ OLAP connectivity or module
- ■ User accessible Data Dictionary
- ■ Internet/Intranet access enabled

Microsoft Technology
- ■ MAPI Compliant
- ■ MFC Used
- ■ ODBC Compliant
- ■ OLE 2/ActiveX Complaint
- □ Designed for Windows 95 Logo
- □ Office Compatible Logo
- □ BackOffice Powered Logo

MARKETING INFORMATION

Start Price Per Module
- □ < US$25K
- ■ < US$50K
- □ < US$100K
- □ > US$100K

Sales Channel
- □ Direct Only
- □ VAR Only
- ■ Direct and VAR
- □ Big Six / Outsource Firm

Direct Sales & Support Offices
- ■ In USA/Canada
- ■ In UK/Europe
- ■ In Pacific Rim/Australasia
- □ In Central/South America

Concepts Dynamic, Inc.

Address Data	**Contact Data**	**Facts and Figures**
Concepts Dynamic Inc.	Tel: (847) 397-4400	Founded: 1981
1821 Walden Office Square, Ste. 500	Fax: (847) 397-0575	Country: USA
Schaumburg, IL 60173	Web: http://www.conceptsdyn.com	Status: Private
		WW User Base: 150
		'95 Revenues ($M): N/A

CDI Financial Control/CDI Project Control

Financials
- ■ General Ledger
- ■ Financial Reporting
- ■ Consolidations
- ■ Budgeting
- ■ Project Management
- ■ Fixed Assets

Supply Chain
- ■ Accounts Payable
- ■ Purchasing
- ■ Accounts Receivable
- ▤ Order Entry   ■ Billing
- ▤ Inventory   ▤ Bill of Materials
- ▤ EDI

Other
- ▤ Manufacturing
- ▤ Human Resources
- ▤ Payroll & Benefits
- ■ Multi Currency
- ■ Multi Lingual
- ■ Allocations

C/S PLATFORM SUPPORT

Server Operating System
- ■ UNIX - IBM
- ■ UNIX - HP
- ■ UNIX - SUN
- ■ Microsoft NT Server
- □ Novell Netware
- □ IBM AS/400 (OS/400)
- □ IBM Mainframe

RDBMS Engine
- □ Oracle 7
- ■ Informix
- □ Sybase 11
- □ Microsoft SQL Server 6.x
- □ Centura SQLBase
- □ BTI Scaleable SQL/Btrieve
- □ IBM DB2 (any platform)

Client GUI
- ■ Windows 95/NT (32 bit)
- ■ Windows 3.x (16 bit)
- □ Apple MACOS
- □ IBM OS/2 Warp (32 bit)
- ■ OSF/Motif
- □ Sun OpenLook
- ■ Character

TECHNOLOGY SUPPORT

Development Technology
- □ C
- □ COBOL
- □ SmallTalk
- □ Progress
- □ PowerBuilder
- □ MS Access
- □ Other 4GL
- □ C++
- □ RPG
- □ Other OO
- ■ NewEra
- □ SQL Windows
- □ Visual Basic

Application Technology
- ■ E mail enabled
- ■ Workflow enabled
- □ Document/image enabled
- □ Telephony enabled
- □ OLAP connectivity or module
- ■ User accessible Data Dictionary
- ■ Internet/Intranet access enabled

Microsoft Technology
- □ MAPI Compliant
- ■ MFC Used
- ■ ODBC Compliant
- ■ OLE 2/ActiveX Complaint
- □ Designed for Windows 95 Logo
- □ Office Compatible Logo
- □ BackOffice Powered Logo

MARKETING INFORMATION

Start Price Per Module
- ■ < US$25K
- □ < US$50K
- □ < US$100K
- □ > US$100K

Sales Channel
- ■ Direct Only
- □ VAR Only
- □ Direct and VAR
- □ Big Six / Outsource Firm

Direct Sales & Support Offices
- ■ In USA/Canada
- □ In UK/Europe
- □ In Pacific Rim/Australasia
- □ In Central/South America

Deltek Systems Inc.

Address Data	Contact Data	Facts and Figures
Deltek Systems Inc.	Tel: (703) 734-8606 x430	Founded: 1983
8280 Greensboro Drive	Fax: (703) 734-1146	Country: USA
McLean VA 22102	Web: http://www.deltek.com	Status: Private
		WW User Base: 1,900
		'96 Revenues ($M): 33

Costpoint

Financials
- ■ General Ledger
- ■ Financial Reporting
- ■ Consolidations
- ■ Budgeting
- ■ Project Management
- ■ Fixed Assets

Supply Chain
- ■ Accounts Payable
- ■ Purchasing
- ■ Accounts Receivable
- ■ Order Entry   ■ Billing
- ■ Inventory   ■ Bill of Materials
- ■ EDI

Other
- ■ Manufacturing
- ■ Human Resources
- ■ Payroll & Benefits
- ■ Multi Currency (1997)
- ☐ Multi Lingual
- ■ Allocations

C/S PLATFORM SUPPORT

Server Operating System
- ■ UNIX - IBM
- ■ UNIX - HP
- ■ UNIX - SUN
- ■ Microsoft NT Server
- ■ Novell Netware
- ☐ IBM AS/400 (OS/400)
- ☐ IBM Mainframe

RDBMS Engine
- ■ Oracle 7
- ☐ Informix
- ■ Sybase 11
- ■ Microsoft SQL Server 6.x
- ■ Centura SQLBase
- ☐ BTI Scaleable SQL/Btrieve
- ☐ IBM DB2 (any platform)

Client GUI
- ■ Windows 95/NT (32 bit)
- ■ Windows 3.x (16 bit)
- ☐ Apple MACOS
- ☐ IBM OS/2 Warp (32 bit)
- ☐ OSF/Motif
- ☐ Sun OpenLook
- ☐ Character

TECHNOLOGY SUPPORT

Development Technology
- ■ C   ■ C++
- ☐ COBOL   ☐ RPG
- ☐ SmallTalk   ☐ Other OO
- ☐ Progress   ☐ NewEra
- ☐ PowerBuilder ■ SQL Windows
- ☐ MS Access   ☐ Visual Basic
- ☐ Other 4GL

Application Technology
- ■ E mail enabled
- ■ Workflow enabled
- ☐ Document/image enabled
- ☐ Telephony enabled
- ■ OLAP connectivity or module
- ☐ User accessible Data Dictionary
- ■ Internet/Intranet access enabled

Microsoft Technology
- ☐ MAPI Compliant
- ☐ MFC Used
- ■ ODBC Compliant
- ■ OLE 2/ActiveX Complaint
- ☐ Designed for Windows 95 Logo
- ☐ Office Compatible Logo
- ☐ BackOffice Powered Logo

MARKETING INFORMATION

Start Price Per Module
- ■ < US$25K
- ☐ < US$50K
- ☐ < US$100K
- ☐ > US$100K

Sales Channel
- ■ Direct Only
- ☐ VAR Only
- ☐ Direct and VAR
- ☐ Big Six / Outsource Firm

Direct Sales & Support Offices
- ■ In USA/Canada
- ☐ In UK/Europe
- ☐ In Pacific Rim/Australasia
- ☐ In Central/South America

Design Data Systems

Address Data	**Contact Data**		**Facts and Figures**	
Design Data Systems Corp.	Tel:	(813) 539-1077	Founded:	1988
11701 South Blecher Road, Ste. 105	Fax:	(813) 539-8042	Country:	USA
Largo, FL 33775	Web:	http://www.designdatasys.com	WW User Base:	2,200
			'95 Revenues ($M):	3.5

SQL*Time Financials

Financials
- ■ General Ledger
- ■ Financial Reporting
- ■ Consolidations
- ■ Budgeting
- ■ Project Management
- ■ Fixed Assets

Supply Chain
- ■ Accounts Payable
- ■ Purchasing
- ■ Accounts Receivable
- ■ Order Entry   ■ Billing
- ■ Inventory   ■ Bill of Materials
- ☐ EDI

Other
- ☐ Manufacturing
- ☐ Human Resources
- ▣ Payroll & Benefits
- ■ Multi Currency
- ☐ Multi Lingual
- ■ Allocations

C/S PLATFORM SUPPORT

Server Operating System
- ■ UNIX - IBM
- ■ UNIX - HP
- ■ UNIX - SUN
- ■ Microsoft NT Server
- ■ Novell Netware
- ■ IBM AS/400 (OS/400)
- ☐ IBM Mainframe

RDBMS Engine
- ■ Oracle 7
- ☐ Informix
- ☐ Sybase 11
- ☐ Microsoft SQL Server 6.x
- ☐ Centura SQLBase
- ☐ BTI Scaleable SQL/Btrieve
- ☐ IBM DB2 (any platform)

Client GUI
- ■ Windows 95/NT (32 bit)
- ■ Windows 3.x (16 bit)
- ■ Apple MACOS
- ☐ IBM OS/2 Warp (32 bit)
- ■ OSF/Motif
- ☐ Sun OpenLook
- ☐ Character

TECHNOLOGY SUPPORT

Development Technology

☐ C	☐ C++
☐ COBOL	☐ RPG
☐ SmallTalk	☐ Other OO
☐ Progress	☐ NewEra
☐ PowerBuilder	■ SQL Windows
☐ MS Access	☐ Visual Basic
☐ Other 4GL: Triton	

Application Technology
- ■ E mail enabled
- ■ Workflow enabled
- ■ Document/image enabled
- ☐ Telephony enabled
- ■ OLAP connectivity or module
- ☐ User accessible Data Dictionary
- ■ Internet/Intranet access enabled

Microsoft Technology
- ■ MAPI Compliant
- ☐ MFC Used
- ☐ ODBC Compliant
- ■ OLE 2/ActiveX Complaint
- ■ Designed for Windows 95 Logo
- ☐ Office Compatible Logo
- ☐ BackOffice Powered Logo

MARKETING INFORMATION

Start Price Per Module
- ☐ < US$25K
- ■ < US$50K
- ☐ < US$100K
- ☐ > US$100K

Sales Channel
- ■ Direct Only
- ☐ VAR Only
- ■ Direct and VAR
- ■ Big Six / Outsource Firm

Direct Sales & Support Offices
- ■ In USA/Canada
- ☐ In UK/Europe
- ■ In Pacific Rim/Australasia
- ■ In Central/South America

FlexiInternational Software Inc.

Address Data	**Contact Data**	**Facts and Figures**
FlexiInternational Software Inc.	Tel: (203) 925-3040	Founded: 1991
Two Enterprise Drive	Fax: (203) 925-3044	Country: USA
Shelton CT 06484	Web: http://www.flexi.com	Status: Private
		WW User Base: N/A
		'95 Revenues ($M): N/A

FlexiFinancials

Financials	**Supply Chain**	**Other**
■ General Ledger	■ Accounts Payable	☐ Manufacturing
■ Financial Reporting	■ Purchasing	☐ Human Resources
■ Consolidations	■ Accounts Receivable	☐ Payroll & Benefits
■ Budgeting	■ Order Entry ☐ Billing	■ Multi Currency
■ Project Management	■ Inventory ☐ Bill of Materials	■ Multi Lingual
■ Fixed Assets	■ EDI	■ Allocations

C/S PLATFORM SUPPORT

Server Operating System	**RDBMS Engine**	**Client GUI**
■ UNIX - IBM	■ Oracle 7	■ Windows 95/NT (32 bit)
■ UNIX - HP	☐ Informix	■ Windows 3.x (16 bit)
■ UNIX - SUN	■ Sybase 11	☐ Apple MACOS
■ Microsoft NT Server	■ Microsoft SQL Server 6.x	☐ IBM OS/2 Warp (32 bit)
■ Novell Netware	☐ Centura SQLBase	☐ OSF/Motif
■ IBM AS/400 (OS/400)	☐ BTI Scaleable SQL/Btrieve	☐ Sun OpenLook
■ IBM Mainframe	■ IBM DB2 (any platform)	☐ Character

TECHNOLOGY SUPPORT

Development Technology		**Application Technology**	**Microsoft Technology**
■ C	■ C++	■ E mail enabled	■ MAPI Compliant
☐ COBOL	☐ RPG	■ Workflow enabled	■ MFC Used
☐ SmallTalk	☐ Other OO	■ Document/image enabled	■ ODBC Compliant
☐ Progress	☐ NewEra	☐ Telephony enabled	■ OLE 2/ActiveX Complaint
☐ PowerBuilder	☐ SQL Windows	■ OLAP connectivity or module	☐ Designed for Windows 95 Logo
☐ MS Access	☐ Visual Basic	■ User accessible Data Dictionary	☐ Office Compatible Logo
☐ Other 4GL		■ Internet/Intranet access enabled	■ BackOffice Powered Logo

MARKETING INFORMATION

Start Price Per Module	**Sales Channel**	**Direct Sales & Support Offices**
☐ < US$25K	■ Direct Only	■ In USA/Canada
■ < US$50K	☐ VAR Only	■ In UK/Europe
☐ < US$100K	☐ Direct and VAR	■ In Pacific Rim/Australasia
☐ > US$100K	■ Big Six / Outsource Firm	☐ In Central/South America

Fourth Shift Corporation

Address Data
Fourth Shift Corporation
7900 International Drive
Minneapolis MN 55425

Contact Data
Tel: (800) 342-5675
Fax: (612) 851-1560
Web: http://www.fs.com

Facts and Figures
Founded: 1982
Country: USA
Status: Public
WW user Base: 2,500
'95 Revenues ($M): 37.2

Manufacturing Software System (MSS)

Financials
- ■ General Ledger
- ■ Financial Reporting
- ■ Consolidations
- ■ Budgeting
- ☐ Project Management
- ☐ Fixed Assets

Supply Chain
- ■ Accounts Payable
- ■ Purchasing
- ■ Accounts Receivable
- ■ Order Entry ☐ Billing
- ☐ Inventory ■ Bill of Materials
- ■ EDI

Other
- ■ Manufacturing
- ☐ Human Resources
- ☐ Payroll & Benefits
- ■ Multi Currency
- ■ Multi Lingual
- ☐ Allocations

C/S PLATFORM SUPPORT

Server Operating System
- ☐ UNIX - IBM
- ☐ UNIX - HP
- ☐ UNIX - SUN
- ■ Microsoft NT Server
- ■ Novell Netware
- ☐ IBM AS/400 (OS/400)
- ☐ IBM Mainframe

RDBMS Engine
- ☐ Oracle 7
- ☐ Informix
- ☐ Sybase 11
- ■ Microsoft SQL Server 6.x
- ☐ Centura SQLBase
- ☐ BTI Scaleable SQL/Btrieve
- ☐ IBM DB2 (any platform)

Client GUI
- ■ Windows 95/NT (32 bit)
- ■ Windows 3.x (16 bit)
- ☐ Apple MACOS
- ☐ IBM OS/2 Warp (32 bit)
- ☐ OSF/Motif
- ☐ Sun OpenLook
- ■ Character

TECHNOLOGY SUPPORT

Development Technology
- ■ C   ■ C++
- ☐ COBOL   ☐ RPG
- ☐ SmallTalk ☐ Other OO
- ☐ Progress   ☐ NewEra
- ☐ PowerBuilder ☐ SQL Windows
- ☐ MS Access   ☐ Visual Basic
- ☐ Other 4GL: Triton

Application Technology
- ☐ E mail enabled
- ☐ Workflow enabled
- ☐ Document/image enabled
- ☐ Telephony enabled
- ☐ OLAP connectivity or module
- ☐ User accessible Data Dictionary
- ☐ Internet/Intranet access enabled

Microsoft Technology
- ☐ MAPI Compliant
- ☐ MFC Used
- ■ ODBC Compliant
- ☐ OLE 2/ActiveX Complaint
- ☐ Designed for Windows 95 Logo
- ☐ Office Compatible Logo
- ☐ BackOffice Powered Logo

MARKETING INFORMATION

Start Price Per Module
- ■ < US$25K
- ☐ < US$50K
- ☐ < US$100K
- ☐ > US$100K

Sales Channel
- ☐ Direct Only
- ☐ VAR Only
- ■ Direct and VAR
- ☐ Big Six / Outsource Firm

Direct Sales & Support Offices
- ■ In USA/Canada
- ■ In UK/Europe
- ■ In Pacific Rim/Australasia
- ■ In Central/South America

Geac SmartStream

Address Data
Geac SmartStream
66 Perimeter Center East
Atlanta GA 30346

Contact Data
Tel: (404) 239-2000
Fax: (404) 239-4933
Web: http://www.dbsoftware.com

Facts and Figures
Founded:
Country: USA
Status: Private
WW user Base: 4000
'95 Revenues ($M): 360

SmartStream

Financials
- General Ledger
- Financial Reporting
- Consolidations
- Budgeting
- Project Management
- Fixed Assets

Supply Chain
- Accounts Payable
- Purchasing
- Accounts Receivable
- Order Entry ■ Billing
- Inventory ■ Bill of Materials
- EDI

Other
- Manufacturing
- Human Resources
- Payroll & Benefits
- Multi Currency
- Multi Lingual
- Allocations

C/S PLATFORM SUPPORT

Server Operating System
- ■ UNIX - IBM
- ■ UNIX - HP
- ■ UNIX - SUN
- ■ Microsoft NT Server
- ■ Novell Netware
- ☐ IBM AS/400 (OS/400)
- ☐ IBM Mainframe

RDBMS Engine
- ☐ Oracle 7
- ☐ Informix
- ■ Sybase 11
- ■ Microsoft SQL Server 6.x
- ☐ Centura SQLBase
- ☐ BTI Scaleable SQL/Btrieve
- ☐ IBM DB2 (any platform)

Client GUI
- ■ Windows 95/NT (32 bit)
- ■ Windows 3.x (16 bit)
- ☐ Apple MACOS
- ☐ IBM OS/2 Warp (32 bit)
- ☐ OSF/Motif
- ☐ Sun OpenLook
- ☐ Character

TECHNOLOGY SUPPORT

Development Technology
- ☐ C   ■ C++
- ☐ COBOL   ☐ RPG
- ☐ SmallTalk   ☐ Other OO
- ☐ Progress   ☐ NewEra
- ■ PowerBuilder   ☐ SQL Windows
- ■ MS Access   ■ Visual Basic
- ☐ Other 4GL

Application Technology
- ■ E mail enabled
- ■ Workflow enabled
- ■ Document/image enabled
- ■ Telephony enabled
- ■ OLAP connectivity or module
- ☐ User accessible Data Dictionary
- ■ Internet/Intranet access enabled

Microsoft Technology
- ■ MAPI Compliant
- ☐ MFC Used
- ■ ODBC Compliant
- ■ OLE 2/ActiveX Complaint
- ☐ Designed for Windows 95 Logo
- ☐ Office Compatible Logo
- ☐ BackOffice Powered Logo

MARKETING INFORMATION

Start Price Per Module
- ☐ < US$25K
- ☐ < US$50K
- ■ < US$100K
- ☐ > US$100K

Sales Channel
- ☐ Direct Only
- ☐ VAR Only
- ■ Direct and VAR
- ☐ Big Six / Outsource Firm

Direct Sales & Support Offices
- ■ In USA/Canada
- ■ In UK/Europe
- ■ In Pacific Rim/Australasia
- ☐ In Central/South America

GeacVision*Shift*

Address Data	**Contact Data**	**Facts and Figures**
GeacVisionShift	Tel: (813) 872-9990	Founded: 1975
3707 West Cherry Street	Fax: (813) 876-8786	Country: Canada
Tampa FL 33607	Web: http://www.geac.com/vs/	Status: Public
		WW User Base: 1200+
		'96 Revenues ($M): 205

Vision*Shift*

Financials
- ■ General Ledger
- ■ Financial Reporting
- ■ Consolidations
- ■ Budgeting
- ☐ Project Management
- ☐ Fixed Assets

Supply Chain
- ■ Accounts Payable
- ■ Purchasing
- ■ Accounts Receivable
- ☐ Order Entry   ☐ Billing
- ☐ Inventory   ☐ Bill of Materials
- ☐ EDI

Other
- ☐ Manufacturing
- ■ Human Resources
- ■ Payroll & Benefits
- ■ Multi Currency
- ■ Multi Lingual
- ■ Allocations

C/S PLATFORM SUPPORT

Server Operating System
- ☐ UNIX - IBM
- ☐ UNIX - HP
- ☐ UNIX - SUN
- ■ Microsoft NT Server
- ■ Novell Netware
- ☐ IBM AS/400 (OS/400)
- ☐ IBM Mainframe

RDBMS Engine
- ☐ Oracle 7
- ☐ Informix
- ☐ Sybase 11
- ■ Microsoft SQL Server 6.x
- ☐ Centura SQLBase
- ☐ BTI Scaleable SQL/Btrieve
- ☐ IBM DB2 (any platform)

Client GUI
- ■ Windows 95/NT (32 bit)
- ■ Windows 3.x (16 bit)
- ☐ Apple MACOS
- ☐ IBM OS/2 Warp (32 bit)
- ☐ OSF/Motif
- ☐ Sun OpenLook
- ☐ Character

TECHNOLOGY SUPPORT

Development Technology
- ☐ C
- ☐ COBOL
- ☐ SmallTalk
- ☐ Progress
- ☐ PowerBuilder
- ■ MS Access
- ☐ Other 4GL
- ☐ C++
- ☐ RPG
- ☐ Other OO
- ☐ NewEra
- ☐ SQL Windows
- ■ Visual Basic

Application Technology
- ■ E mail enabled
- ■ Workflow enabled
- ■ Document/image enabled
- ☐ Telephony enabled
- ☐ OLAP connectivity or module
- ■ User accessible Data Dictionary
- ☐ Internet/Intranet access enabled

Microsoft Technology
- ■ MAPI Compliant
- ☐ MFC Used
- ■ ODBC Compliant
- ■ OLE 2/ActiveX Complaint
- ■ Designed for Windows 95 Logo
- ■ Office Compatible Logo
- ☐ BackOffice Powered Logo

MARKETING INFORMATION

Start Price Per Module
- ■ < US$25K
- ☐ < US$50K
- ☐ < US$100K
- ☐ > US$100K

Sales Channel
- ☐ Direct Only
- ☐ VAR Only
- ■ Direct and VAR
- ☐ Big Six / Outsource Firm

Direct Sales & Support Offices
- ■ In USA/Canada
- ☐ In UK/Europe
- ■ In Pacific Rim/Australasia
- ☐ In Central/South America

Great Plains Software

Address Data	**Contact Data**	**Facts and Figures**
Great Plains Software	Tel: (701) 281-0550	Founded: 1981
1701 SW 38th St.	Fax: (701) 281-3700	Country: USA
Fargo, ND 58103	Web: http://www.gps.com	Status: Private
		WW User Base: 500
		'95 Revenues ($M): N/A

Dynamics C/S+

Financials	**Supply Chain**	**Other**
■ General Ledger	■ Accounts Payable	▣ Manufacturing
■ Financial Reporting	■ Purchasing	▣ Human Resources
■ Consolidations	■ Accounts Receivable	■ Payroll & Benefits
■ Budgeting	■ Order Entry ■ Billing	■ Multi Currency
☐ Project Management	■ Inventory ▣ Bill of Materials	■ Multi Lingual
■ Fixed Assets	▣ EDI	■ Allocations

C/S PLATFORM SUPPORT

Server Operating System	**RDBMS Engine**	**Client GUI**
☐ UNIX - IBM	☐ Oracle 7	■ Windows 95/NT (32 bit)
☐ UNIX - HP	☐ Informix	■ Windows 3.x (16 bit)
☐ UNIX - SUN	☐ Sybase 11	☐ Apple MACOS
■ Microsoft NT Server	■ Microsoft SQL Server 6.x	☐ IBM OS/2 Warp (32 bit)
☐ Novell Netware	☐ Centura SQLBase	☐ OSF/Motif
☐ IBM AS/400 (OS/400)	■ BTI Scaleable SQL/Btrieve	☐ Sun OpenLook
☐ IBM Mainframe	☐ IBM DB2 (any platform)	☐ Character

TECHNOLOGY SUPPORT

Development Technology		**Application Technology**	**Microsoft Technology**
■ C	■ C++	■ E mail enabled	■ MAPI Compliant
☐ COBOL	☐ RPG	■ Workflow enabled	■ MFC Used
☐ SmallTalk	☐ Other OO	▣ Document/image enabled	■ ODBC Compliant
☐ Progress	☐ NewEra	☐ Telephony enabled	■ OLE 2/ActiveX Complaint
☐ PowerBuilder	☐ SQL Windows	☐ OLAP connectivity or module	■ Designed for Windows 95 Logo
☐ MS Access	☐ Visual Basic	■ User accessible Data Dictionary	☐ Office Compatible Logo
■ Other 4GL: GPS Dexterity		■ Internet/Intranet access enabled	■ BackOffice Powered Logo

MARKETING INFORMATION

Start Price Per Module	**Sales Channel**	**Direct Sales & Support Offices**
■ < US$25K	☐ Direct Only	☐ In USA/Canada
☐ < US$50K	■ VAR Only	☐ In UK/Europe
☐ < US$100K	☐ Direct and VAR	☐ In Pacific Rim/Australasia
☐ > US$100K	■ Big Six / Outsource Firm	☐ In Central/South America

Hyperion Software Corp.

Address Data	**Contact Data**	**Facts and Figures**
Hyperion Software Corp.	Tel: (203) 703-3000	Founded: 1981
900 Long Ridge Road	Fax: (203) 595-8500	Country: USA
Stamford CT 06902	Web: http://www.hysoft.com	Status: Public
		WW User Base: 3100+
		'96 Revenues ($M): 194

Hyperion Solutions

Financials
- ■ General Ledger
- ■ Financial Reporting
- ■ Consolidations
- ■ Budgeting
- □ Project Management
- ■ Fixed Assets

Supply Chain
- ■ Accounts Payable
- ■ Purchasing
- ■ Accounts Receivable
- □ Order Entry   □ Billing
- □ Inventory   □ Bill of Materials
- ▣ EDI

Other
- □ Manufacturing
- □ Human Resources
- □ Payroll & Benefits
- ■ Multi Currency
- ■ Multi Lingual
- ■ Allocations

C/S PLATFORM SUPPORT

Server Operating System
- ■ UNIX - IBM
- ■ UNIX - HP
- ■ UNIX - SUN
- ■ Microsoft NT Server
- □ Novell Netware
- □ IBM AS/400 (OS/400)
- □ IBM Mainframe

RDBMS Engine
- ■ Oracle 7
- □ Informix
- ■ Sybase 11
- □ Microsoft SQL Server 6.x
- □ Centura SQLBase
- □ BTI Scaleable SQL/Btrieve
- □ IBM DB2 (any platform)

Client GUI
- ■ Windows 95/NT (32 bit)
- ■ Windows 3.x (16 bit)
- □ Apple MACOS
- □ IBM OS/2 Warp (32 bit)
- □ OSF/Motif
- □ Sun OpenLook
- □ Character

TECHNOLOGY SUPPORT

Development Technology
- ■ C
- □ COBOL
- □ SmallTalk
- □ Progress
- □ PowerBuilder
- ■ MS Access
- ■ C++
- □ RPG
- □ Other OO
- □ NewEra
- □ SQL Windows
- ■ Visual Basic

Application Technology
- □ E mail enabled
- ■ Workflow enabled
- □ Document/image enabled
- □ Telephony enabled
- ■ OLAP connectivity or module
- □ User accessible Data Dictionary
- ■ Internet/Intranet access enabled

Microsoft Technology
- □ MAPI Compliant
- ■ MFC Used
- ■ ODBC Compliant
- ■ OLE 2/ActiveX Complaint
- □ Designed for Windows 95 Logo
- □ Office Compatible Logo
- □ BackOffice Powered Logo

MARKETING INFORMATION

Start Price Per Module
- □ < US$25K
- □ < US$50K
- ■ < US$100K
- ■ > US$100K

Sales Channel
- ■ Direct Only
- □ VAR Only
- □ Direct and VAR
- □ Big Six / Outsource Firm

Direct Sales & Support Offices
- ■ In USA/Canada
- ■ In UK/Europe
- ■ In Pacific Rim/Australasia
- ■ In Central/South America

IET, Inc.

Address Data
IET, Inc.
4001 S. 700 E., Ste. 301
Salt Lake City UT 84107

Contact Data
Tel: (801) 281-6600
Fax: (801) 284-9160
Web: www.iet.xmission.com

Facts and Figures
Founded: 1995
Country: USA
Status: Private
WW User Base: N/A
'95 Revenues ($M): N/A

Perspectives

Financials
- ■ General Ledger
- ■ Financial Reporting
- ☐ Consolidations
- ■ Budgeting
- ■ Project Management
- ■ Fixed Assets

Supply Chain
- ■ Accounts Payable
- ■ Purchasing
- ■ Accounts Receivable
- ■ Order Entry   ■ Billing
- ■ Inventory   ☐ Bill of Materials
- ☐ EDI

Other
- ▥ Manufacturing
- ■ Human Resources
- ■ Payroll & Benefits
- ■ Multi Currency (planned)
- ■ Multi Lingual (planned)
- ■ Allocations

C/S PLATFORM SUPPORT

Server Operating System
- ■ UNIX - IBM
- ■ UNIX - HP
- ■ UNIX - SUN
- ■ Microsoft NT Server
- ☐ Novell Netware
- ☐ IBM AS/400 (OS/400)
- ☐ IBM Mainframe

RDBMS Engine
- ■ Oracle 7
- ■ Informix
- ■ Sybase 11
- ■ Microsoft SQL Server 6.x
- ☐ Centura SQLBase
- ☐ BTI Scaleable SQL/Btrieve
- ■ IBM DB2 (any platform) (planned)

Client GUI
- ■ Windows 95/NT (32 bit)
- ■ Windows 3.x (16 bit)
- ☐ Apple MACOS
- ☐ IBM OS/2 Warp (32 bit)
- ☐ OSF/Motif
- ☐ Sun OpenLook
- ☐ Character

TECHNOLOGY SUPPORT

Development Technology
- ☐ C
- ☐ COBOL
- ☐ SmallTalk
- ☐ Progress
- ☐ PowerBuilder
- ☐ MS Access
- ☐ Other 4GL
- ■ C++
- ☐ RPG
- ☐ Other OO
- ☐ NewEra
- ☐ SQL Windows
- ☐ Visual Basic

Application Technology
- ■ E mail enabled
- ☐ Workflow enabled
- ☐ Document/image enabled
- ☐ Telephony enabled
- ☐ OLAP connectivity or module
- ☐ User accessible Data Dictionary
- ■ Internet/Intranet access enabled (planned)

Microsoft Technology
- ■ MAPI Compliant
- ☐ MFC Used
- ■ ODBC Compliant
- ■ OLE 2/ActiveX Complaint
- ☐ Designed for Windows 95 Logo
- ☐ Office Compatible Logo
- ☐ BackOffice Powered Logo

MARKETING INFORMATION

Start Price Per Module
- ☐ Please contact vendor

Sales Channel
- ■ Direct Only
- ☐ VAR Only
- ☐ Direct and VAR
- ☐ Big Six / Outsource Firm

Direct Sales & Support Offices
- ■ In USA/Canada
- ☐ In UK/Europe
- ☐ In Pacific Rim/Australasia
- ☐ In Central/South America

JBA International

Address Data
JBA International
3701 Algonquin Road
Rolling Meadows IL 60008

Contact Data
Tel: (800) JBA-INTL
Fax: (847) 590-0049
Web: http://www.jbaintl.com

Facts and Figures
Founced: 1981
Country: UK
Status: Public
WW user Base:
'95 Revenues ($M): N/A

System 21

Financials
■ General Ledger
■ Financial Reporting
■ Consolidations
■ Budgeting
■ Project Management
■ Fixed Assets

Supply Chain
■ Accounts Payable
■ Purchasing
■ Accounts Receivable
■ Order Entry ■ Billing
■ Inventory ■ Bill of Materials
■ EDI

Other
■ Manufacturing
☐ Human Resources
☐ Payroll & Benefits
■ Multi Currency
■ Multi Lingual
■ Allocations

C/S PLATFORM SUPPORT

Server Operating System
■ UNIX - IBM
■ UNIX - HP
☐ UNIX - SUN
■ Microsoft NT Server
☐ Novell Netware
■ IBM AS/400 (OS/400)
☐ IBM Mainframe

RDBMS Engine
■ Oracle 7
☐ Informix
☐ Sybase 11
☐ Microsoft SQL Server 6.x
☐ Centura SQLBase
☐ BTI Scaleable SQL/Btrieve
■ IBM DB2 (any platform)

Client GUI
■ Windows 95/NT (32 bit)
■ Windows 3.x (16 bit)
☐ Apple MACOS
☐ IBM OS/2 Warp (32 bit)
☐ OSF/Motif
☐ Sun OpenLook
☐ Character

TECHNOLOGY SUPPORT

Development Technology
☐ C   ■ C++
☐ COBOL   ■ RPG
☐ SmallTalk ☐ Other OO
☐ Progress   ☐ NewEra
☐ PowerBuilder ☐ SQL Windows
☐ MS Access ■ Visual Basic
■ Other 4GL: JOT ■ Internet/Intranet access enabled

Application Technology
■ E mail enabled
■ Workflow enabled
■ Document/image enabled
■ Telephony enabled
■ OLAP connectivity or module
■ User accessible Data Dictionary

Microsoft Technology
☐ MAPI Compliant
☐ MFC Used
■ ODBC Compliant
■ OLE 2/ActiveX Complaint
■ Designed for Windows 95 Logo
■ Office Compatible Logo
☐ BackOffice Powered Logo

MARKETING INFORMATION

Start Price Per Module
■ < US$25K
☐ < US$50K
☐ < US$100K
☐ > US$100K

Sales Channel
☐ Direct Only
☐ VAR Only
■ Direct and VAR
☐ Big Six / Outsource Firm

Direct Sales & Support Offices
■ In USA/Canada
■ In UK/Europe
■ In Pacific Rim/Australasia
■ In Central/South America

J D Edwards

Address Data
J D Edwards
8055 East Tufts, Ste. 1500
Denver CO 80215

Contact Data
Tel: (303) 488-4000
Fax: (303) 488-4842
Web: http://www.jdedwards.com

Facts and Figures
Founded: 1977
Country: USA
Status: Private
WW User Base: 3518
'95 Revenues ($M): 341

OneWorld

Financials
- ■ General Ledger
- ■ Financial Reporting
- ■ Consolidations
- ■ Budgeting
- ■ Project Management
- ■ Fixed Assets

Supply Chain
- ■ Accounts Payable
- ■ Purchasing
- ■ Accounts Receivable
- ■ Order Entry   ■ Billing
- ■ Inventory   ■ Bill of Materials
- ■ EDI

Other
- ■ Manufacturing
- ■ Human Resources
- ■ Payroll & Benefits
- ■ Multi Currency
- ■ Multi Lingual
- ■ Allocations

C/S PLATFORM SUPPORT

Server Operating System
- ■ UNIX - IBM
- ■ UNIX - HP
- □ UNIX - SUN
- ■ Microsoft NT Server
- □ Novell Netware
- ■ IBM AS/400 (OS/400)
- ■ IBM Mainframe

RDBMS Engine
- ■ Oracle 7
- □ Informix
- □ Sybase 11
- ■ Microsoft SQL Server 6.x
- □ Centura SQLBase
- □ BTI Scaleable SQL/Btrieve
- ■ IBM DB2 (any platform)

Client GUI
- ■ Windows 95/NT (32 bit)
- □ Windows 3.x (16 bit)
- □ Apple MACOS
- □ IBM OS/2 Warp (32 bit)
- □ OSF/Motif
- □ Sun OpenLook
- ■ Character

TECHNOLOGY SUPPORT

Development Technology
- ■ C   ■ C++
- □ COBOL   ■ RPG
- □ SmallTalk   □ Other OO
- □ Progress   □ NewEra
- □ PowerBuilder □ SQL Windows
- □ MS Access   □ Visual Basic
- □ Other 4GL

Application Technology
- ■ E mail enabled
- ■ Workflow enabled
- ■ Document/image enabled
- □ Telephony enabled
- ■ OLAP connectivity or module
- ■ User accessible Data Dictionary
- ■ Internet/Intranet access enabled

Microsoft Technology
- □ MAPI Compliant
- □ MFC Used
- ■ ODBC Compliant
- ■ OLE 2/ActiveX Complaint
- □ Designed for Windows 95 Logo
- □ Office Compatible Logo
- □ BackOffice Powered Logo

MARKETING INFORMATION

Start Price Per Module
- □ < US$25K
- □ < US$50K
- □ < US$100K
- ■ > US$100K

Sales Channel
- ■ Direct Only
- □ VAR Only
- ■ Direct and VAR
- □ Big Six / Outsource Firm

Direct Sales & Support Offices
- ■ In USA/Canada
- ■ In UK/Europe
- ■ In Pacific Rim/Australasia
- ■ In Central/South America

Lawson Software

Address Data	**Contact Data**	**Facts and Figures**
Lawson Software	Tel: (800) 477-1357	Founded: 1975
1300 Godward St.	Fax: (612) 379-7141	Country: USA
Minneapolis MN 55413	Web: http://www.lawson.com	Status: Private
		WW User Base: N/A
		'95 Revenues ($M): 101.8

Insight

Financials
- ■ General Ledger
- ■ Financial Reporting
- ■ Consolidations
- ■ Budgeting
- ■ Project Management
- ■ Fixed Assets

Supply Chain
- ■ Accounts Payable
- ■ Purchasing
- ■ Accounts Receivable
- ■ Order Entry ■ Billing
- ■ Inventory ■ Bill of Materials
- ■ EDI

Other
- ☐ Manufacturing
- ■ Human Resources
- ■ Payroll & Benefits
- ■ Multi Currency
- ■ Multi Lingual
- ■ Allocations

C/S PLATFORM SUPPORT

Server Operating System
- ■ UNIX - IBM
- ■ UNIX - HP
- ■ UNIX - SUN
- ■ Microsoft NT Server
- ☐ Novell Netware
- ■ IBM AS/400 (OS/400)
- ☐ IBM Mainframe

RDBMS Engine
- ■ Oracle 7
- ■ Informix
- ■ Sybase 11
- ☐ Microsoft SQL Server 6.x
- ☐ Centura SQLBase
- ☐ BTI Scaleable SQL/Btrieve
- ☐ IBM DB2 (any platform)

Client GUI
- ■ Windows 95/NT (32 bit)
- ■ Windows 3.x (16 bit)
- ☐ Apple MACOS
- ☐ IBM OS/2 Warp (32 bit)
- ☐ OSF/Motif
- ☐ Sun OpenLook
- ■ Character

TECHNOLOGY SUPPORT

Development Technology
- ■ C
- ■ COBOL
- ☐ SmallTalk
- ☐ Progress
- ☐ PowerBuilder
- ☐ MS Access
- ■ Other 4GL: Lawson 4GL
- ■ C++
- ■ RPG
- ☐ Other OO
- ☐ NewEra
- ☐ SQL Windows
- ■ Visual Basic

Application Technology
- ■ E mail enabled
- ■ Workflow enabled
- ■ Document/image enabled
- ☐ Telephony enabled
- ■ OLAP connectivity or module
- ■ User accessible Data Dictionary
- ■ Internet/Intranet access enabled

Microsoft Technology
- ■ MAPI Compliant
- ☐ MFC Used
- ■ ODBC Compliant
- ■ OLE 2/ActiveX Complaint
- ☐ Designed for Windows 95 Logo
- ☐ Office Compatible Logo
- ☐ BackOffice Powered Logo

MARKETING INFORMATION

Start Price Per Module
- ■ < US$25K
- ■ < US$50K
- ☐ < US$100K
- ☐ > US$100K

Sales Channel
- ☐ Direct Only
- ☐ VAR Only
- ■ Direct and VAR
- ☐ Big Six / Outsource Firm

Direct Sales & Support Offices
- ■ In USA/Canada
- ■ In UK/Europe
- ☐ In Pacific Rim/Australasia
- ☐ In Central/South America

Macola Inc.

Address Data	**Contact Data**	**Facts and Figures**
Macola Inc.	Tel: (614) 382-5999	Founded: 1971
333 E. Center St.	Fax: (614) 382-0239	Country: USA
Marion OH 43302	Web: http://www.macola.com	Status: Private
		WW User Base: N/A
		'95 Revenues ($M): N/A

Progression Series

Financials
- ■ General Ledger
- ■ Financial Reporting
- ■ Consolidations
- ■ Budgeting
- ■ Project Management
- ■ Fixed Assets

Supply Chain
- ■ Accounts Payable
- ■ Purchasing
- ■ Accounts Receivable
- ■ Order Entry □ Billing
- ■ Inventory ■ Bill of Materials
- ■ EDI

Other
- ■ Manufacturing
- ■ Human Resources
- ■ Payroll & Benefits
- ■ Multi Currency
- ■ Multi Lingual
- □ Allocations

C/S PLATFORM SUPPORT

Server Operating System
- □ UNIX - IBM
- □ UNIX - HP
- □ UNIX - SUN
- ■ Microsoft NT Server
- ■ Novell Netware
- □ IBM AS/400 (OS/400)
- □ IBM Mainframe

RDBMS Engine
- □ Oracle 7
- □ Informix
- □ Sybase 11
- □ Microsoft SQL Server 6.x
- □ Centura SQLBase
- ■ BTI Scaleable SQL/Btrieve
- □ IBM DB2 (any platform)

Client GUI
- ■ Windows 95/NT (32 bit)
- ■ Windows 3.x (16 bit)
- □ Apple MACOS
- □ IBM OS/2 Warp (32 bit)
- □ OSF/Motif
- □ Sun OpenLook
- □ Character

TECHNOLOGY SUPPORT

Development Technology
- □ C ■ C++
- ■ COBOL □ RPG
- □ SmallTalk □ Other OO
- □ Progress □ NewEra
- □ PowerBuilder □ SQL Windows
- □ MS Access ■ Visual Basic
- □ Other 4GL

Application Technology
- □ E mail enabled
- □ Workflow enabled
- □ Document/image enabled
- □ Telephony enabled
- □ OLAP connectivity or module
- ■ User accessible Data Dictionary
- □ Internet/Intranet access enabled

Microsoft Technology
- □ MAPI Compliant
- □ MFC Used
- ■ ODBC Compliant
- □ OLE 2/ActiveX Complaint
- □ Designed for Windows 95 Logo
- □ Office Compatible Logo
- □ BackOffice Powered Logo

MARKETING INFORMATION

Start Price Per Module
- ■ < US$25K
- □ < US$50K
- □ < US$100K
- □ > US$100K

Sales Channel
- □ Direct Only
- ■ VAR Only
- □ Direct and VAR
- □ Big Six / Outsource Firm

Direct Sales & Support Offices
- ■ In USA/Canada
- ■ In UK/Europe
- □ In Pacific Rim/Australasia
- □ In Central/South America

Maconomy NE Inc.

Address Data
Maconomy NE Inc.
124 Anderson Road
Marlborough MA 01752

Contact Data
Tel: (508) 460-8337
Fax: (508) 460-6327
Web: http://www.maconomy-usa.com

Facts and Figures
Founded: 1989
Country: Denmark
Status: Private
WW User Base: 4000
'95 Revenues ($M): N/A

Maconomy Job Cost

Financials
■ General Ledger
■ Financial Reporting
■ Consolidations
■ Budgeting
■ Project Management
■ Fixed Assets

Supply Chain
■ Accounts Payable
■ Purchasing
■ Accounts Receivable
■ Order Entry   ■ Billing
■ Inventory   ■ Bill of Materials
■ EDI

Other
■ Manufacturing
☐ Human Resources
☐ Payroll & Benefits
■ Multi Currency
■ Multi Lingual
■ Allocations

C/S PLATFORM SUPPORT

Server Operating System
■ UNIX - IBM
■ UNIX - HP
■ UNIX - SUN
■ Microsoft NT Server
☐ Novell Netware
☐ IBM AS/400 (OS/400)
☐ IBM Mainframe

RDBMS Engine
■ Oracle 7
☐ Informix
■ Sybase 11
☐ Microsoft SQL Server 6.x
☐ Centura SQLBase
☐ BTI Scaleable SQL/Btrieve
☐ IBM DB2 (any platform)

Client GUI
■ Windows 95/NT (32 bit)
■ Windows 3.x (16 bit)
■ Apple MACOS
■ IBM OS/2 Warp (32 bit)
☐ OSF/Motif
☐ Sun OpenLook
☐ Character

TECHNOLOGY SUPPORT

Development Technology
☐ C   ☐ C++
☐ COBOL   ☐ RPG
☐ SmallTalk   ☐ Other OO
☐ Progress   ☐ NewEra
☐ PowerBuilder ☐ SQL Windows
☐ MS Access   ☐ Visual Basic
■ Other 4GL: Maconomy 4GL

Application Technology
■ E mail enabled
■ Workflow enabled
■ Document/image enabled
☐ Telephony enabled
■ OLAP connectivity or module
■ User accessible Data Dictionary
■ Internet/Intranet access enabled

Microsoft Technology
☐ MAPI Compliant
☐ MFC Used
■ ODBC Compliant
☐ OLE 2/ActiveX Complaint
☐ Designed for Windows 95 Logo
☐ Office Compatible Logo
☐ BackOffice Powered Logo

MARKETING INFORMATION

Start Price Per Module
☐ < US$25K
■ < US$50K
☐ < US$100K
☐ > US$100K

Sales Channel
☐ Direct Only
☐ VAR Only
■ Direct and VAR
☐ Big Six / Outsource Firm

Direct Sales & Support Offices
■ In USA/Canada
■ In UK/Europe
☐ In Pacific Rim/Australasia
☐ In Central/South America

MTX International Inc.

Address Data	**Contact Data**	**Facts and Figures**
MTX International Inc.	Tel: (303) 770-9840	Founded: 1976
98 Inverness Drive E., Ste. 110	Fax: (303) 770-6711	Country: USA
Englewood CO 80112	Web: http://www.mtxi.com	Status: Public
		WW User Base: 2000
		'95 Revenues ($M): 1

MTX Accounting for Microsoft Office

Financials
- ■ General Ledger
- ■ Financial Reporting
- ■ Consolidations
- ■ Budgeting
- ☐ Project Management
- ▣ Fixed Assets

Supply Chain
- ■ Accounts Payable
- ■ Purchasing
- ■ Accounts Receivable
- ■ Order Entry   ■ Billing
- ■ Inventory   ☐ Bill of Materials
- ☐ EDI

Other
- ▣ Manufacturing
- ▣ Human Resources
- ■ Payroll & Benefits
- ☐ Multi Currency
- ☐ Multi Lingual
- ☐ Allocations

C/S PLATFORM SUPPORT

Server Operating System
- ☐ UNIX - IBM
- ☐ UNIX - HP
- ☐ UNIX - SUN
- ■ Microsoft NT Server
- ■ Novell Netware
- ☐ IBM AS/400 (OS/400)
- ☐ IBM Mainframe

RDBMS Engine
- ☐ Oracle 7
- ☐ Informix
- ☐ Sybase 11
- ■ Microsoft SQL Server 6.x
- ☐ Centura SQLBase
- ☐ BTI Scaleable SQL/Btrieve
- ☐ IBM DB2 (any platform)

Client GUI
- ■ Windows 95/NT (32 bit)
- ■ Windows 3.x (16 bit)
- ☐ Apple MACOS
- ☐ IBM OS/2 Warp (32 bit)
- ☐ OSF/Motif
- ☐ Sun OpenLook
- ☐ Character

TECHNOLOGY SUPPORT

Development Technology
- ☐ C   ☐ C++
- ☐ COBOL   ☐ RPG
- ☐ SmallTalk   ☐ Other OO
- ☐ Progress   ☐ NewEra
- ☐ PowerBuilder ☐ SQL Windows
- ■ MS Access   ☐ Visual Basic
- ☐ Other 4GL

Application Technology
- ■ E mail enabled
- ☐ Workflow enabled
- ☐ Document/image enabled
- ☐ Telephony enabled
- ☐ OLAP connectivity or module
- ■ User accessible Data Dictionary
- ■ Internet/Intranet access enabled

Microsoft Technology
- ■ MAPI Compliant
- ☐ MFC Used
- ■ ODBC Compliant
- ■ OLE 2/ActiveX Complaint
- ■ Designed for Windows 95 Logo
- ■ Office Compatible Logo
- ■ BackOffice Powered Logo

MARKETING INFORMATION

Start Price Per Module
- ■ < US$25K
- ☐ < US$50K
- ☐ < US$100K
- ☐ > US$100K

Sales Channel
- ☐ Direct Only
- ☐ VAR Only
- ■ Direct and VAR
- ☐ Big Six / Outsource Firm

Direct Sales & Support Offices
- ■ In USA/Canada
- ☐ In UK/Europe
- ☐ In Pacific Rim/Australasia
- ☐ In Central/South America

Navision Software US Inc.

Address Data	**Contact Data**	**Facts and Figures**
Navision Software US Inc.	Tel: (770) 564-8014	Founded: 1987
One Meca Way	Fax: (770) 564-8010	Country: Denmark
Norcross GA 30093	Web: http://www.navision-us.com	Status: Private
		WW User Base: 25,000
		'95 Revenues ($M): N/A

Navision Financials

Financials
- ■ General Ledger
- ■ Financial Reporting
- ■ Consolidations
- ■ Budgeting
- ■ Project Management
- ▣ Fixed Assets

Supply Chain
- ■ Accounts Payable
- ■ Purchasing
- ■ Accounts Receivable
- ■ Order Entry   ■ Billing
- ■ Inventory   ■ Bill of Materials
- ■ EDI

Other
- ☐ Manufacturing
- ■ Human Resources
- ■ Payroll & Benefits
- ■ Multi Currency
- ■ Multi Lingual
- ■ Allocations

C/S PLATFORM SUPPORT

Server Operating System
- ■ UNIX - IBM
- ■ UNIX - HP
- ☐ UNIX - SUN
- ■ Microsoft NT Server
- ☐ Novell Netware
- ☐ IBM AS/400 (OS/400)
- ☐ IBM Mainframe

RDBMS Engine
- ☐ Oracle 7
- ☐ Informix
- ☐ Sybase 11
- ☐ Microsoft SQL Server 6.x
- ☐ Centura SQLBase
- ☐ BTI Scaleable SQL/Btrieve
- ☐ IBM DB2 (any platform)

Client GUI
- ■ Windows 95/NT (32 bit)
- ■ Windows 3.x (16 bit)
- ☐ Apple MACOS
- ■ IBM OS/2 Warp (32 bit)
- ☐ OSF/Motif
- ☐ Sun OpenLook
- ☐ Character

TECHNOLOGY SUPPORT

Development Technology
- ☐ C
- ☐ COBOL
- ☐ SmallTalk
- ☐ Progress
- ☐ PowerBuilder
- ☐ MS Access
- ■ Other 4GL: Navision C/SIDE
- ■ C++
- ☐ RPG
- ☐ Other OO
- ☐ NewEra
- ☐ SQL Windows
- ☐ Visual Basic

Application Technology
- ☐ E mail enabled
- ☐ Workflow enabled
- ■ Document/image enabled
- ☐ Telephony enabled
- ☐ OLAP connectivity or module
- ☐ User accessible Data Dictionary
- ■ Internet/Intranet access enabled

Microsoft Technology
- ☐ MAPI Compliant
- ☐ MFC Used
- ☐ ODBC Compliant
- ☐ OLE 2/ActiveX Complaint
- ■ Designed for Windows 95 Logo
- ■ Office Compatible Logo
- ■ BackOffice Powered Logo

MARKETING INFORMATION

Start Price Per Module
- ■ < US$25K
- ☐ < US$50K
- ☐ < US$100K
- ☐ > US$100K

Sales Channel
- ☐ Direct Only
- ■ VAR Only
- ☐ Direct and VAR
- ☐ Big Six / Outsource Firm

Direct Sales & Support Offices
- ■ In USA/Canada
- ■ In UK/Europe
- ☐ In Pacific Rim/Australasia
- ☐ In Central/South America

OpenPlus International Inc.

Address Data
OpenPlus International Inc.
3925 West Braker Lane, Ste. 305
Austin TX 78759

Contact Data
Tel:　(512) 328-1231
Fax:　(512) 328-1491
Web:　http://www.openplus.com

Facts and Figures
Founded:　1992 (Legacy: 1972)
Country:　Australia
Status:　Private
WW User Base:　300+
'95 Revenues ($M):　N/A

OpenPlus Enterprise

Financials
- ■ General Ledger
- ■ Financial Reporting
- ■ Consolidations
- ■ Budgeting
- ■ Project Management
- ■ Fixed Assets

Supply Chain
- ■ Accounts Payable
- ■ Purchasing
- ■ Accounts Receivable
- ■ Order Entry　■ Billing
- ■ Inventory　■ Bill of Materials
- ■ EDI

Other
- ☐ Manufacturing
- ■ Human Resources
- ■ Payroll & Benefits
- ■ Multi Currency
- ■ Multi Lingual
- ■ Allocations

C/S PLATFORM SUPPORT

Server Operating System
- ■ UNIX - IBM
- ■ UNIX - HP
- ■ UNIX - SUN
- ■ Microsoft NT Server
- ■ Novell Netware
- ■ IBM AS/400 (OS/400)
- ■ IBM Mainframe

RDBMS Engine
- ■ Oracle 7
- ■ Informix
- ■ Sybase 11
- ■ Microsoft SQL Server 6.x
- ☐ Centura SQLBase
- ☐ BTI Scaleable SQL/Btrieve
- ■ IBM DB2 (any platform)

Client GUI
- ■ Windows 95/NT (32 bit)
- ■ Windows 3.x (16 bit)
- ■ Apple MACOS
- ■ IBM OS/2 Warp (32 bit)
- ■ OSF/Motif
- ■ Sun OpenLook
- ■ Character

TECHNOLOGY SUPPORT

Development Technology
- ☐ C　　　☐ C++
- ☐ COBOL　☐ RPG
- ☐ SmallTalk　☐ Other OO
- ☐ Progress　☐ NewEra
- ☐ PowerBuilder　☐ SQL Windows
- ☐ MS Access　☐ Visual Basic
- ■ Other 4GL: Uniface

Application Technology
- ■ E mail enabled
- ■ Workflow enabled
- ■ Document/image enabled
- ☐ Telephony enabled
- ■ OLAP connectivity or module
- ■ User accessible Data Dictionary
- ■ Internet/Intranet access enabled

Microsoft Technology
- ■ MAPI Compliant
- ☐ MFC Used
- ■ ODBC Compliant
- ■ OLE 2/ActiveX Complaint
- ■ Designed for Windows 95 Logo
- ■ Office Compatible Logo
- ■ BackOffice Powered Logo

MARKETING INFORMATION

Start Price Per Module
- ☐ < US$25K
- ☐ < US$50K
- ■ < US$100K
- ☐ > US$100K

Sales Channel
- ☐ Direct Only
- ☐ VAR Only
- ■ Direct and VAR
- ■ Big Six / Outsource Firm

Direct Sales & Support Offices
- ■ In USA/Canada
- ■ In UK/Europe
- ■ In Pacific Rim/Australasia
- ☐ In Central/South America

Open Systems Inc.

Address Data	**Contact Data**	**Facts and Figures**
Open Systems Inc.	Tel: (800) 328-2276	Founded: 1976
7626 Golden Triangle Drive	Fax: (612) 829-1493	Country: USA
Eden Prairie MN 55347	Web: http://www.osas.com	Status: Private
		WW User Base: N/A
		'95 Revenues ($M): N/A

Traverse

Financials
- ■ General Ledger
- ■ Financial Reporting
- ■ Consolidations
- ■ Budgeting
- ■ Project Management
- ☐ Fixed Assets

Supply Chain
- ■ Accounts Payable
- ■ Purchasing
- ■ Accounts Receivable
- ■ Order Entry   ■ Billing
- ■ Inventory   ■ Bill of Materials
- ☐ EDI

Other
- ☐ Manufacturing
- ☐ Human Resources
- ■ Payroll & Benefits
- ■ Multi Currency
- ■ Multi Lingual
- ■ Allocations

C/S PLATFORM SUPPORT

Server Operating System
- ☐ UNIX - IBM
- ☐ UNIX - HP
- ☐ UNIX - SUN
- ■ Microsoft NT Server
- ■ Novell Netware
- ☐ IBM AS/400 (OS/400)
- ☐ IBM Mainframe

RDBMS Engine
- ☐ Oracle 7
- ☐ Informix
- ☐ Sybase 11
- ■ Microsoft SQL Server 6.x
- ☐ Centura SQLBase
- ☐ BTI Scaleable SQL/Btrieve
- ☐ IBM DB2 (any platform)

Client GUI
- ■ Windows 95/NT (32 bit)
- ■ Windows 3.x (16 bit)
- ☐ Apple MACOS
- ☐ IBM OS/2 Warp (32 bit)
- ☐ OSF/Motif
- ☐ Sun OpenLook
- ☐ Character

TECHNOLOGY SUPPORT

Development Technology
- ☐ C
- ☐ COBOL
- ☐ SmallTalk
- ☐ Progress
- ☐ PowerBuilder
- ■ MS Access
- ☐ Other 4GL
- ☐ C++
- ☐ RPG
- ☐ Other OO
- ☐ NewEra
- ☐ SQL Windows
- ■ Visual Basic

Application Technology
- ■ E mail enabled
- ■ Workflow enabled
- ☐ Document/image enabled
- ☐ Telephony enabled
- ☐ OLAP connectivity or module
- ☐ User accessible Data Dictionary
- ☐ Internet/Intranet access enabled

Microsoft Technology
- ■ MAPI Compliant
- ■ MFC Used
- ■ ODBC Compliant
- ■ OLE 2/ActiveX Complaint
- ■ Designed for Windows 95 Logo
- ☐ Office Compatible Logo
- ☐ BackOffice Powered Logo

MARKETING INFORMATION

Start Price Per Module
- ■ < US$25K
- ☐ < US$50K
- ☐ < US$100K
- ☐ > US$100K

Sales Channel
- ☐ Direct Only
- ■ VAR Only
- ☐ Direct and VAR
- ☐ Big Six / Outsource Firm

Direct Sales & Support Offices
- ■ In USA/Canada
- ■ In UK/Europe
- ■ In Pacific Rim/Australasia
- ☐ In Central/South America

Orange Systems, ALCIE Financials Division

Address Data
Orange Systems
13577 Feather Sound Drive #1450
Clearwater FL 34622-5539

Contact Data
Tel: (813) 571-1606 x241
Fax: (813) 571-1703
Web: http://www.alcie.com

Facts and Figures
Founded: 1971
Country: USA
Status: Private
WW User Base: 1000
'95 Revenues ($M): 30

ALCIE

Financials
- ■ General Ledger
- ■ Financial Reporting
- ■ Consolidations
- ■ Budgeting
- ■ Project Management
- ■ Fixed Assets

Supply Chain
- ■ Accounts Payable
- ■ Purchasing
- ■ Accounts Receivable
- ■ Order Entry ■ Billing
- ■ Inventory ■ Bill of Materials
- ☐ EDI

Other
- ■ Manufacturing
- ☐ Human Resources
- ■ Payroll & Benefits
- ☐ Multi Currency
- ☐ Multi Lingual
- ■ Allocations

C/S PLATFORM SUPPORT

Server Operating System
- ■ UNIX - IBM
- ■ UNIX - HP
- ■ UNIX - SUN
- ■ Microsoft NT Server
- ■ Novell Netware
- ☐ IBM AS/400 (OS/400)
- ☐ IBM Mainframe

RDBMS Engine
- ■ Oracle 7
- ☐ Informix
- ☐ Sybase 11
- ☐ Microsoft SQL Server 6.x
- ☐ Centura SQLBase
- ☐ BTI Scaleable SQL/Btrieve
- ☐ IBM DB2 (any platform)

Client GUI
- ☐ Windows 95/NT (32 bit)
- ☐ Windows 3.x (16 bit)
- ☐ Apple MACOS
- ☐ IBM OS/2 Warp (32 bit)
- ☐ OSF/Motif
- ☐ Sun OpenLook
- ■ Character

TECHNOLOGY SUPPORT

Development Technology
- ☐ C
- ☐ COBOL
- ☐ SmallTalk
- ☐ Progress
- ☐ PowerBuilder
- ☐ MS Access
- ■ Other 4GL: Oracle Forms
- ☐ C++
- ☐ RPG
- ☐ Other OO
- ☐ NewEra
- ☐ SQL Windows
- ☐ Visual Basic

Application Technology
- ☐ E mail enabled
- ☐ Workflow enabled
- ☐ Document/image enabled
- ☐ Telephony enabled
- ☐ OLAP connectivity or module
- ■ User accessible Data Dictionary
- ☐ Internet/Intranet access enabled

Microsoft Technology
- ☐ MAPI Compliant
- ☐ MFC Used
- ☐ ODBC Compliant
- ☐ OLE 2/ActiveX Complaint
- ☐ Designed for Windows 95 Logo
- ☐ Office Compatible Logo
- ☐ BackOffice Powered Logo

MARKETING INFORMATION

Start Price Per Module
- ■ < US$25K
- ☐ < US$50K
- ☐ < US$100K
- ☐ > US$100K

Sales Channel
- ☐ Direct Only
- ☐ VAR Only
- ■ Direct and VAR
- ☐ Big Six / Outsource Firm

Direct Sales & Support Offices
- ■ In USA/Canada
- ☐ In UK/Europe
- ☐ In Pacific Rim/Australasia
- ☐ In Central/South America

PeopleSoft

Address Data
PeopleSoft
4440 Rosewood Drive
Pleasanton CA 94588-3031

Contact Data
Tel: (800) 947-7753
Fax: (510) 468-2399
Web: http://www.peoplesoft.com

Facts and Figures
Founded: 1987
Country: USA
Status: Public
WW User Base: 1300+
'96 Revenues ($M): 450

PeopleSoft

Financials
- General Ledger
- Financial Reporting
- Consolidations
- Budgeting
- Project Management
- Fixed Assets

Supply Chain
- Accounts Payable
- Purchasing
- Accounts Receivable
- Order Entry
- Billing
- Inventory
- Bill of Materials
- EDI

Other
- Manufacturing
- Human Resources
- Payroll & Benefits
- Multi Currency
- Multi Lingual
- Allocations

C/S PLATFORM SUPPORT

Server Operating System
- UNIX - IBM
- UNIX - HP
- UNIX - SUN
- Microsoft NT Server
- Novell Netware
- IBM AS/400 (OS/400)
- IBM Mainframe

RDBMS Engine
- Oracle 7
- Informix
- Sybase 11
- Microsoft SQL Server 6.x
- Centura SQLBase
- ☐ BTI Scaleable SQL/Btrieve
- IBM DB2 (any platform)

Client GUI
- Windows 95/NT (32 bit)
- Windows 3.x (16 bit)
- ☐ Apple MACOS
- ☐ IBM OS/2 Warp (32 bit)
- ☐ OSF/Motif
- ☐ Sun OpenLook
- ☐ Character

TECHNOLOGY SUPPORT

Development Technology
- ☐ C
- ☐ COBOL
- ☐ SmallTalk
- ☐ Progress
- ☐ PowerBuilder
- ☐ MS Access
- Other 4GL: PeopleTools
- ☐ C++
- ☐ RPG
- ☐ Other OO
- ☐ NewEra
- ☐ SQL Windows
- ☐ Visual Basic

Application Technology
- E mail enabled
- Workflow enabled
- Document/image enabled
- Telephony enabled
- OLAP connectivity or module
- User accessible Data Dictionary
- Internet/Intranet access enabled

Microsoft Technology
- ☐ MAPI Compliant
- ☐ MFC Used
- ODBC Compliant
- ☐ OLE 2/ActiveX Complaint
- ☐ Designed for Windows 95 Logo
- ☐ Office Compatible Logo
- ☐ BackOffice Powered Logo

MARKETING INFORMATION

Start Price Per Module
- ☐ < US$25K
- ☐ < US$50K
- < US$100K
- ☐ > US$100K

Sales Channel
- Direct Only
- ☐ VAR Only
- ☐ Direct and VAR
- ☐ Big Six / Outsource Firm

Direct Sales & Support Offices
- In USA/Canada
- In UK/Europe
- In Pacific Rim/Australasia
- In Central/South America

Platinum Software Corporation

Address Data
Platinum Software Corporation
195 Technology Drive
Irvine CA 92618

Contact Data
Tel: (800) 426-0469
Fax: (714) 727-4005
Web: http://www.platsoft.com

Facts and Figures
Founded: 1984
Country: USA
Status: Public
WW User Base: 40,000
'96 Revenues ($M):   41

Platinum SQL

Financials
- ■ General Ledger
- ■ Financial Reporting
- ■ Consolidations
- ■ Budgeting
- □ Project Management
- ■ Fixed Assets

Supply Chain
- ■ Accounts Payable
- ■ Purchasing
- ■ Accounts Receivable
- ■ Order Entry   ■ Billing
- ■ Inventory   □ Bill of Materials
- ▣ EDI

Other
- ▣ Manufacturing
- ▣ Human Resources
- ▣ Payroll & Benefits
- ■ Multi Currency
- ■ Multi Lingual
- ■ Allocations

C/S PLATFORM SUPPORT

Server Operating System
- □ UNIX - IBM
- ■ UNIX - HP
- ■ UNIX - SUN
- ■ Microsoft NT Server
- □ Novell Netware
- □ IBM AS/400 (OS/400)
- □ IBM Mainframe

RDBMS Engine
- □ Oracle 7
- □ Informix
- ■ Sybase 11
- ■ Microsoft SQL Server 6.x
- □ Centura SQLBase
- □ BTI Scaleable SQL/Btrieve
- □ IBM DB2 (any platform)

Client GUI
- ■ Windows 95/NT (32 bit)
- ■ Windows 3.x (16 bit)
- □ Apple MACOS
- □ IBM OS/2 Warp (32 bit)
- □ OSF/Motif
- □ Sun OpenLook
- □ Character

TECHNOLOGY SUPPORT

Development Technology
- □ C   ■ C++
- □ COBOL   □ RPG
- □ SmallTalk   □ Other OO
- □ Progress   □ NewEra
- □ PowerBuilder □ SQL Windows
- □ MS Access   ■ Visual Basic
- □ Other 4GL

Application Technology
- ■ E mail enabled
- □ Workflow enabled
- □ Document/image enabled
- □ Telephony enabled
- ■ OLAP connectivity or module
- ■ User accessible Data Dictionary
- ■ Internet/Intranet access enabled

Microsoft Technology
- ■ MAPI Compliant
- ■ MFC Used
- ■ ODBC Compliant
- ■ OLE 2/ActiveX Complaint
- □ Designed for Windows 95 Logo
- □ Office Compatible Logo
- ■ BackOffice Powered Logo

MARKETING INFORMATION

Start Price Per Module
- ■ < US$25K
- □ < US$50K
- □ < US$100K
- □ > US$100K

Sales Channel
- □ Direct Only
- □ VAR Only
- ■ Direct and VAR
- □ Big Six / Outsource Firm

Direct Sales & Support Offices
- ■ In USA/Canada
- ■ In UK/Europe
- ■ In Pacific Rim/Australasia
- ■ In Central/South America

PowerCerv

Address Data	Contact Data	Facts and Figures
PowerCerv	Tel: (813) 226-2100	Founded: 1992
400 N. Ashley	Fax: (813) 222-0886	Country: USA
Tampa FL 33602	Web: http://www.platsoft.com	Status: Public
		WW User Base: 10,000
		'96 Revenues ($M): 28.2

ADAPTlication for Financials

Financials
- ■ General Ledger
- ■ Financial Reporting
- ■ Consolidations
- □ Budgeting
- ■ Project Management
- ■ Fixed Assets

Supply Chain
- ■ Accounts Payable
- ■ Purchasing
- ■ Accounts Receivable
- ■ Order Entry ■ Billing
- ■ Inventory ■ Bill of Materials
- ■ EDI

Other
- ■ Manufacturing
- □ Human Resources
- □ Payroll & Benefits
- ■ Multi Currency
- □ Multi Lingual
- ■ Allocations

C/S PLATFORM SUPPORT

Server Operating System
- ■ UNIX - IBM
- ■ UNIX - HP
- ■ UNIX - SUN
- ■ Microsoft NT Server
- □ Novell Netware
- □ IBM AS/400 (OS/400)
- □ IBM Mainframe

RDBMS Engine
- ■ Oracle 7
- □ Informix
- □ Sybase 11
- ■ Microsoft SQL Server 6.x
- □ Centura SQLBase
- □ BTI Scaleable SQL/Btrieve
- □ IBM DB2 (any platform)

Client GUI
- ■ Windows 95/NT (32 bit)
- ■ Windows 3.x (16 bit)
- □ Apple MACOS
- □ IBM OS/2 Warp (32 bit)
- □ OSF/Motif
- □ Sun OpenLook
- □ Character

TECHNOLOGY SUPPORT

Development Technology
- □ C
- □ COBOL
- □ SmallTalk
- □ Progress
- ■ PowerBuilder
- □ MS Access
- □ Other 4GL
- □ C++
- □ RPG
- □ Other OO
- □ NewEra
- □ SQL Windows
- ■ Visual Basic

Application Technology
- ■ E mail enabled
- ■ Workflow enabled
- ■ Document/image enabled
- □ Telephony enabled
- □ OLAP connectivity or module
- □ User accessible Data Dictionary
- ■ Internet/Intranet access enabled

Microsoft Technology
- ■ MAPI Compliant
- □ MFC Used
- ■ ODBC Compliant
- □ OLE 2/ActiveX Complaint
- □ Designed for Windows 95 Logo
- □ Office Compatible Logo
- □ BackOffice Powered Logo

MARKETING INFORMATION

Start Price Per Module
- ■ < US$25K
- □ < US$50K
- □ < US$100K
- □ > US$100K

Sales Channel
- □ Direct Only
- □ VAR Only
- ■ Direct and VAR
- □ Big Six / Outsource Firm

Direct Sales & Support Offices
- ■ In USA/Canada
- ■ In UK/Europe
- □ In Pacific Rim/Australasia
- □ In Central/South America

Errata

Page 174 of this text was printed incorrectly.

The correct version of the page is printed

on the reverse side of this sheet.

PowerCerv

Address Data	Contact Data	Facts and Figures
PowerCerv	Tel: (813) 226-2600	Founded: 1992
400 N. Ashley Drive, Ste. 2700	Fax: (813) 222-0886	Country: USA
Tampa FL 33602	Web: http://www.powercerv.com	Status: Public
		WW User Base: 10,000
		Accounting Base: 200
		'96 Revenues ($M): 28.2

ADAPTlication for Financials

Financials	Supply Chain	Other
■ General Ledger	■ Accounts Payable	■ Manufacturing
■ Financial Reporting	■ Purchasing	□ Human Resources
■ Consolidations	■ Accounts Receivable	□ Payroll & Benefits
□ Budgeting	■ Order Entry ■ Billing	■ Multi Currency
■ Project Management	■ Inventory ■ Bill of Materials	□ Multi Lingual
■ Fixed Assets	■ EDI	■ Allocations

C/S PLATFORM SUPPORT

Server Operating System	RDBMS Engine	Client GUI
■ UNIX - IBM	■ Oracle 7	■ Windows 95/NT (32 bit)
■ UNIX - HP	□ Informix	■ Windows 3.x (16 bit)
■ UNIX - SUN	■ Sybase 11	□ Apple MACOS
■ Microsoft NT Server	■ Microsoft SQL Server 6.x	□ IBM OS/2 Warp (32 bit)
□ Novell Netware	□ Centura SQLBase	□ OSF/Motif
□ IBM AS/400 (OS/400)	□ BTI Scaleable SQL/Btrieve	□ Sun OpenLook
□ IBM Mainframe	□ IBM DB2 (any platform)	□ Character

TECHNOLOGY SUPPORT

Development Technology		Application Technology	Microsoft Technology
□ C	□ C++	■ E mail enabled	■ MAPI Compliant
□ COBOL	□ RPG	■ Workflow enabled	□ MFC Used
□ SmallTalk	□ Other OO	■ Document/image enabled	■ ODBC Compliant
□ Progress	□ NewEra	□ Telephony enabled	□ OLE 2/ActiveX Complaint
■ PowerBuilder	□ SQL Windows	□ OLAP connectivity or module	□ Designed for Windows 95 Logo
□ MS Access	■ Visual Basic	□ User accessible Data Dictionary	□ Office Compatible Logo
□ Other 4GL		■ Internet/Intranet access enabled	□ BackOffice Powered Logo

MARKETING INFORMATION

Start Price Per Module	Sales Channel	Direct Sales & Support Offices
■ < US$25K	□ Direct Only	■ In USA/Canada
□ < US$50K	□ VAR Only	■ In UK/Europe
□ < US$100K	■ Direct and VAR	□ In Pacific Rim/Australasia
□ > US$100K	□ Big Six / Outsource Firm	□ In Central/South America

QSP Inc.

Address Data	**Contact Data**	**Facts and Figures**
QSP Inc.	Tel: (919) 872-4100	Founded: 1981
3200 Atlantic Avenue	Fax: (919) 872-4015	Country: UK
Raleigh NC 27604	Web: http://www.qspinc.com	Status: Public
		WW User Base: N/A
		'95 Revenues ($M): 34

Universal OLAS

Financials
- ■ General Ledger
- ■ Financial Reporting
- □ Consolidations
- ■ Budgeting
- □ Project Management
- ■ Fixed Assets

Supply Chain
- ■ Accounts Payable
- ■ Purchasing
- ■ Accounts Receivable
- □ Order Entry   □ Billing
- □ Inventory   □ Bill of Materials
- □ EDI

Other
- □ Manufacturing
- □ Human Resources
- □ Payroll & Benefits
- ■ Multi Currency
- ■ Multi Lingual
- □ Allocations

C/S PLATFORM SUPPORT

Server Operating System
- ■ UNIX - IBM
- ■ UNIX - HP
- ■ UNIX - SUN
- □ Microsoft NT Server
- □ Novell Netware
- □ IBM AS/400 (OS/400)
- ■ IBM Mainframe

RDBMS Engine
- ■ Oracle 7
- □ Informix
- □ Sybase 11
- □ Microsoft SQL Server 6.x
- □ Centura SQLBase
- □ BTI Scaleable SQL/Btrieve
- ■ IBM DB2 (any platform)

Client GUI
- □ Windows 95/NT (32 bit)
- ■ Windows 3.x (16 bit)
- □ Apple MACOS
- ■ IBM OS/2 Warp (32 bit)
- □ OSF/Motif
- □ Sun OpenLook
- ■ Character

TECHNOLOGY SUPPORT

Development Technology
- ■ C   □ C++
- ■ COBOL   □ RPG
- □ SmallTalk   □ Other OO
- □ Progress   □ NewEra
- □ PowerBuilder □ SQL Windows
- □ MS Access   ■ Visual Basic
- □ Other 4GL

Application Technology
- ■ E mail enabled
- ■ Workflow enabled
- ■ Document/image enabled
- □ Telephony enabled
- ■ OLAP connectivity or module
- □ User accessible Data Dictionary
- □ Internet/Intranet access enabled

Microsoft Technology
- □ MAPI Compliant
- □ MFC Used
- ■ ODBC Compliant
- □ OLE 2/ActiveX Complaint
- □ Designed for Windows 95 Logo
- □ Office Compatible Logo
- □ BackOffice Powered Logo

MARKETING INFORMATION

Start Price Per Module
- □ < US$25K
- □ < US$50K
- ■ < US$100K
- □ > US$100K

Sales Channel
- ■ Direct Only
- □ VAR Only
- □ Direct and VAR
- □ Big Six / Outsource Firm

Direct Sales & Support Offices
- ■ In USA/Canada
- ■ In UK/Europe
- □ In Pacific Rim/Australasia
- □ In Central/South America

Ramco Systems Corp.

Address Data	Contact Data	Facts and Figures
Address Data Ramco Systems Corp. 2201 Walnut Avenue Fremont CA 94538	**Contact Data** Tel: (510) 494-2964 Fax: (510) 494-2979 Web: http://www.ramco.com	**Facts and Figures** Founded: 1989 Country: India Status: Private WW User Base: N/A '95 Revenues ($M): N/A

Marshal

Financials
- ■ General Ledger
- ■ Financial Reporting
- ■ Consolidations
- ■ Budgeting
- □ Project Management
- ■ Fixed Assets

Supply Chain
- ■ Accounts Payable
- ■ Purchasing
- ■ Accounts Receivable
- ■ Order Entry   ■ Billing
- ■ Inventory   ■ Bill of Materials
- ■ EDI

Other
- ■ Manufacturing
- ■ Human Resources
- □ Payroll & Benefits
- ■ Multi Currency
- ■ Multi Lingual
- ■ Allocations

C/S PLATFORM SUPPORT

Server Operating System
- □ UNIX - IBM
- □ UNIX - HP
- □ UNIX - SUN
- ■ Microsoft NT Server
- □ Novell Netware
- □ IBM AS/400 (OS/400)
- □ IBM Mainframe

RDBMS Engine
- □ Oracle 7
- □ Informix
- □ Sybase 11
- ■ Microsoft SQL Server 6.x
- □ Centura SQLBase
- □ BTI Scaleable SQL/Btrieve
- □ IBM DB2 (any platform)

Client GUI
- ■ Windows 95/NT (32 bit)
- ■ Windows 3.x (16 bit)
- □ Apple MACOS
- □ IBM OS/2 Warp (32 bit)
- □ OSF/Motif
- □ Sun OpenLook
- □ Character

TECHNOLOGY SUPPORT

Development Technology
- □ C   ■ C++
- □ COBOL   □ RPG
- □ SmallTalk   □ Other OO
- □ Progress   □ NewEra
- □ PowerBuilder □ SQL Windows
- □ MS Access □ Visual Basic
- □ Other 4GL

Application Technology
- ■ E mail enabled
- ■ Workflow enabled
- □ Document/image enabled
- □ Telephony enabled
- □ OLAP connectivity or module
- □ User accessible Data Dictionary
- □ Internet/Intranet access enabled

Microsoft Technology
- ■ MAPI Compliant
- ■ MFC Used
- ■ ODBC Compliant
- ■ OLE 2/ActiveX Complaint
- □ Designed for Windows 95 Logo
- □ Office Compatible Logo
- □ BackOffice Powered Logo

MARKETING INFORMATION

Start Price Per Module
- □ < US$25K
- ■ < US$50K
- □ < US$100K
- □ > US$100K

Sales Channel
- ■ Direct Only
- □ VAR Only
- □ Direct and VAR
- □ Big Six / Outsource Firm

Direct Sales & Support Offices
- ■ In USA/Canada
- ■ In UK/Europe
- ■ In Pacific Rim/Australasia
- □ In Central/South America

Ross Systems Inc.

Address Data
Ross Systems Inc.
1100 Johnson Ferry Road, Ste. 750
Atlanta GA 30342

Contact Data
Tel: (404) 851-1872
Fax: (404) 257-0434
Web: http://www.rossinc.com

Facts and Figures
Founded: 1972
Country: USA
Status: Public
WW User Base: 2700
'95 Revenues ($M): 71

Renaissance CS

Financials
- ■ General Ledger
- ■ Financial Reporting
- ■ Consolidations
- ■ Budgeting
- ■ Project Management
- ■ Fixed Assets

Supply Chain
- ■ Accounts Payable
- ■ Purchasing
- ■ Accounts Receivable
- ■ Order Entry   ■ Billing
- ■ Inventory   □ Bill of Materials
- ■ EDI

Other
- ■ Manufacturing
- ■ Human Resources
- ■ Payroll & Benefits
- ■ Multi Currency
- ■ Multi Lingual
- ■ Allocations

C/S PLATFORM SUPPORT

Server Operating System
- ■ UNIX - IBM
- ■ UNIX - HP
- □ UNIX - SUN
- ■ Microsoft NT Server
- □ Novell Netware
- □ IBM AS/400 (OS/400)
- □ IBM Mainframe

RDBMS Engine
- ■ Oracle 7
- □ Informix
- ■ Sybase 11
- □ Microsoft SQL Server 6.x
- □ Centura SQLBase
- □ BTI Scaleable SQL/Btrieve
- □ IBM DB2 (any platform)

Client GUI
- ■ Windows 95/NT (32 bit)
- ■ Windows 3.x (16 bit)
- □ Apple MACOS
- □ IBM OS/2 Warp (32 bit)
- □ OSF/Motif
- □ Sun OpenLook
- ■ Character

TECHNOLOGY SUPPORT

Development Technology
- □ C   □ C++
- □ COBOL   □ RPG
- □ SmallTalk   □ Other OO
- □ Progress   □ NewEra
- □ PowerBuilder   □ SQL Windows
- □ MS Access   □ Visual Basic
- ■ Other 4GL: Gembase

Application Technology
- ■ E mail enabled
- ■ Workflow enabled
- ■ Document/image enabled
- □ Telephony enabled
- ▣ OLAP connectivity or module
- ■ User accessible Data Dictionary
- □ Internet/Intranet access enabled

Microsoft Technology
- □ MAPI Compliant
- ■ MFC Used
- □ ODBC Compliant
- □ OLE 2/ActiveX Complaint
- □ Designed for Windows 95 Logo
- □ Office Compatible Logo
- □ BackOffice Powered Logo

MARKETING INFORMATION

Start Price Per Module
- ■ < US$25K
- ■ < US$50K
- ■ < US$100K
- ■ > US$100K

Sales Channel
- □ Direct Only
- □ VAR Only
- ■ Direct and VAR
- □ Big Six / Outsource Firm

Direct Sales & Support Offices
- ■ In USA/Canada
- ■ In UK/Europe
- ■ In Pacific Rim/Australasia
- □ In Central/South America

SAP America Inc.

Address Data
SAP America Inc.
701 Lee Road
Wayne PA 19087

Contact Data
Tel: (610) 725-4500
Fax: (610) 725-4555
Web: http://www.sap.com

Facts and Figures
Founded: 1972
Country: Germany
WW User Base: 8000
'95 Revenues ($M): 1961

R/3

Financials
- ■ General Ledger
- ■ Financial Reporting
- ■ Consolidations
- ■ Budgeting
- ■ Project Management
- ■ Fixed Assets

Supply Chain
- ■ Accounts Payable
- ■ Purchasing
- ■ Accounts Receivable
- ■ Order Entry ■ Billing
- ■ Inventory   ■ Bill of Materials
- ■ EDI

Other
- ■ Manufacturing
- ■ Human Resources
- ■ Payroll & Benefits
- ■ Multi Currency
- ■ Multi Lingual
- ■ Allocations

C/S PLATFORM SUPPORT

Server Operating System
- ■ UNIX - IBM
- ■ UNIX - HP
- ■ UNIX - SUN
- ■ Microsoft NT Server
- ☐ Novell Netware
- ■ IBM AS/400 (OS/400)
- ☐ IBM Mainframe

RDBMS Engine
- ■ Oracle 7
- ■ Informix
- ☐ Sybase 11
- ■ Microsoft SQL Server 6.x
- ☐ Centura SQLBase
- ☐ BTI Scaleable SQL/Btrieve
- ■ IBM DB2 (any platform)

Client GUI
- ■ Windows 95/NT (32 bit)
- ■ Windows 3.x (16 bit)
- ■ Apple MACOS
- ■ IBM OS/2 Warp (32 bit)
- ■ OSF/Motif
- ☐ Sun OpenLook
- ☐ Character

TECHNOLOGY SUPPORT

Development Technology
- ■ C
- ☐ COBOL
- ☐ SmallTalk
- ☐ Progress
- ☐ PowerBuilder
- ☐ MS Access
- ■ C++
- ☐ RPG
- ☐ Other OO
- ☐ NewEra
- ☐ SQL Windows
- ☐ Visual Basic
- ■ Other 4GL: ABAP/4

Application Technology
- ■ E mail enabled
- ■ Workflow enabled
- ■ Document/image enabled
- ■ Telephony enabled
- ☐ OLAP connectivity or module
- ■ User accessible Data Dictionary
- ■ Internet/Intranet access enabled

Microsoft Technology
- ■ MAPI Compliant
- ☐ MFC Used
- ■ ODBC Compliant
- ☐ OLE 2/ActiveX Complaint
- ☐ Designed for Windows 95 Logo
- ☐ Office Compatible Logo
- ☐ BackOffice Powered Logo

MARKETING INFORMATION

Start Price Per Module
- ☐ < US$25K
- ☐ < US$50K
- ☐ < US$100K
- ☐ > US$100K

Sales Channel
- ☐ Direct Only
- ☐ VAR Only
- ■ Direct and VAR
- ☐ Big Six / Outsource Firm

Direct Sales & Support Offices
- ■ In USA/Canada
- ■ In UK/Europe
- ■ In Pacific Rim/Australasia
- ■ In Central/South America

SAS Institute Inc.

Address Data	**Contact Data**	**Facts and Figures**
SAS Institute Inc.	Tel: (919) 677-8000	Founded: 1976
SAS Campus Drive	Fax: (919) 677-4444	Country: USA
Cary NC 27511	Web: http://www.sas.com/vision	Status: Private
		WW User Base: 3M
		'95 Revenues ($M): 562

CFO Vision

Financials
- ☐ General Ledger
- ■ Financial Reporting
- ■ Consolidations
- ☐ Budgeting
- ☐ Project Management
- ☐ Fixed Assets

Supply Chain
- ☐ Accounts Payable
- ☐ Purchasing
- ☐ Accounts Receivable
- ☐ Order Entry   ☐ Billing
- ☐ Inventory   ☐ Bill of Materials
- ☐ EDI

Other
- ☐ Manufacturing
- ☐ Human Resources
- ☐ Payroll & Benefits
- ■ Multi Currency
- ☐ Multi Lingual
- ☐ Allocations

C/S PLATFORM SUPPORT

Server Operating System
- ■ UNIX - IBM
- ■ UNIX - HP
- ■ UNIX - SUN
- ■ Microsoft NT Server
- ■ Novell Netware
- ☐ IBM AS/400 (OS/400)
- ☐ IBM Mainframe

RDBMS Engine
- ■ Oracle 7
- ■ Informix
- ■ Sybase 11
- ■ Microsoft SQL Server 6.x
- ☐ Centura SQLBase
- ■ BTI Scaleable SQL/Btrieve
- ■ IBM DB2 (any platform)

Client GUI
- ■ Windows 95/NT (32 bit)
- ■ Windows 3.x (16 bit)
- ☐ Apple MACOS
- ■ IBM OS/2 Warp (32 bit)
- ☐ OSF/Motif
- ☐ Sun OpenLook
- ☐ Character

TECHNOLOGY SUPPORT

Development Technology
- ■ C   ☐ C++
- ☐ COBOL   ☐ RPG
- ☐ SmallTalk   ■ Other OO
- ☐ Progress   ☐ NewEra
- ☐ PowerBuilder   ☐ SQL Windows
- ☐ MS Access   ☐ Visual Basic
- ■ Other 4GL

Application Technology
- ■ E mail enabled
- ☐ Workflow enabled
- ☐ Document/image enabled
- ☐ Telephony enabled
- ■ OLAP connectivity or module
- ☐ User accessible Data Dictionary
- ☐ Internet/Intranet access enabled

Microsoft Technology
- ■ MAPI Compliant
- ☐ MFC Used
- ■ ODBC Compliant
- ■ OLE 2/ActiveX Complaint
- ■ Designed for Windows 95 Logo
- ☐ Office Compatible Logo
- ☐ BackOffice Powered Logo

MARKETING INFORMATION

Start Price Per Module
- ☐ < US$25K
- ☐ < US$50K
- ■ < US$100K
- ☐ > US$100K

Sales Channel
- ■ Direct Only
- ☐ VAR Only
- ☐ Direct and VAR
- ☐ Big Six / Outsource Firm

Direct Sales & Support Offices
- ■ In USA/Canada
- ■ In UK/Europe
- ■ In Pacific Rim/Australasia
- ■ In Central/South America

SBT Accounting Systems

Address Data	**Contact Data**	**Facts and Figures**
SBT Accounting Systems	Tel: (415) 444-9900	Founded: 1980
1401 Los Gamos	Fax: (415) 444-9901	Country: USA
San Rafael CA 94903	Web: http://www.sbt.com/vision	Status: Private
		WW User Base: 3M
		'95 Revenues ($M): 562

SBT Executive Series

Financials
- ■ General Ledger
- ■ Financial Reporting
- ■ Consolidations
- ■ Budgeting
- ■ Project Management
- ■ Fixed Assets

Supply Chain
- ■ Accounts Payable
- ■ Purchasing
- ■ Accounts Receivable
- ■ Order Entry ■ Billing
- ■ Inventory ■ Bill of Materials
- □ EDI

Other
- ▣ Manufacturing
- □ Human Resources
- □ Payroll & Benefits
- ■ Multi Currency
- □ Multi Lingual
- ■ Allocations

C/S PLATFORM SUPPORT

Server Operating System
- □ UNIX - IBM
- □ UNIX - HP
- □ UNIX - SUN
- ■ Microsoft NT Server
- □ Novell Netware
- □ IBM AS/400 (OS/400)
- □ IBM Mainframe

RDBMS Engine
- ■ Oracle 7
- ■ Informix
- ■ Sybase 11
- ■ Microsoft SQL Server 6.x
- □ Centura SQLBase
- ■ BTI Scaleable SQL/Btrieve
- ■ IBM DB2 (any platform)

Client GUI
- ■ Windows 95/NT (32 bit)
- □ Windows 3.x (16 bit)
- □ Apple MACOS
- ■ IBM OS/2 Warp (32 bit)
- □ OSF/Motif
- □ Sun OpenLook
- □ Character

TECHNOLOGY SUPPORT

Development Technology
- □ C
- □ COBOL
- □ SmallTalk
- □ Progress
- ■ PowerBuilder
- □ MS Access
- □ Other 4GL
- □ C++
- □ RPG
- □ Other OO
- □ NewEra
- □ SQL Windows
- □ Visual Basic

Application Technology
- ■ E mail enabled
- ■ Workflow enabled
- □ Document/image enabled
- □ Telephony enabled
- □ OLAP connectivity or module
- □ User accessible Data Dictionary
- ■ Internet/Intranet access enabled

Microsoft Technology
- □ MAPI Compliant
- □ MFC Used
- ■ ODBC Compliant
- □ OLE 2/ActiveX Complaint
- □ Designed for Windows 95 Logo
- □ Office Compatible Logo
- □ BackOffice Powered Logo

MARKETING INFORMATION

Start Price Per Module
- ■ < US$25K
- □ < US$50K
- □ < US$100K
- □ > US$100K

Sales Channel
- □ Direct Only
- ■ VAR Only
- □ Direct and VAR
- □ Big Six / Outsource Firm

Direct Sales & Support Offices
- ■ In USA/Canada
- ■ In UK/Europe
- ■ In Pacific Rim/Australasia
- ■ In Central/South America

Scala North America Inc.

Address Data
Scala North America Inc.
601 South Lake Destiny Road
Ste. 200
Maitland FL 32751

Contact Data
Tel: (407) 875-6999
Fax: (407) 875-9957
Web: http://www.scala-na.com

Facts and Figures
Founded: 1978
Country: Sweden
Status: Public
WW User Base: 14000
'95 Revenues ($M): 100

SCALA

Financials
- ■ General Ledger
- ■ Financial Reporting
- ■ Consolidations
- ■ Budgeting
- ■ Project Management
- ■ Fixed Assets

Supply Chain
- ■ Accounts Payable
- ■ Purchasing
- ■ Accounts Receivable
- ■ Order Entry   ■ Billing
- ■ Inventory   ■ Bill of Materials
- ■ EDI

Other
- ■ Manufacturing
- □ Human Resources
- □ Payroll & Benefits
- ■ Multi Currency
- ■ Multi Lingual
- ■ Allocations

C/S PLATFORM SUPPORT

Server Operating System
- □ UNIX - IBM
- ■ UNIX - HP
- ■ UNIX - SUN
- ■ Microsoft NT Server
- ■ Novell Netware
- □ IBM AS/400 (OS/400)
- □ IBM Mainframe

RDBMS Engine
- □ Oracle 7
- ■ Informix
- □ Sybase 11
- □ Microsoft SQL Server 6.x
- □ Centura SQLBase
- ■ BTI Scaleable SQL/Btrieve
- □ IBM DB2 (any platform)

Client GUI
- ■ Windows 95/NT (32 bit)
- ■ Windows 3.x (16 bit)
- □ Apple MACOS
- □ IBM OS/2 Warp (32 bit)
- □ OSF/Motif
- □ Sun OpenLook
- ■ Character

TECHNOLOGY SUPPORT

Development Technology
- □ C   ■ C++
- □ COBOL   □ RPG
- □ SmallTalk   □ Other OO
- □ Progress   □ NewEra
- □ PowerBuilder   □ SQL Windows
- □ MS Access   □ Visual Basic
- ■ Other 4GL: Business Basic

Application Technology
- □ E mail enabled
- □ Workflow enabled
- □ Document/image enabled
- □ Telephony enabled
- □ OLAP connectivity or module
- □ User accessible Data Dictionary
- □ Internet/Intranet access enabled

Microsoft Technology
- □ MAPI Compliant
- □ MFC Used
- □ ODBC Compliant
- □ OLE 2/ActiveX Complaint
- □ Designed for Windows 95 Logo
- □ Office Compatible Logo
- □ BackOffice Powered Logo

MARKETING INFORMATION

Start Price Per Module
- ■ < US$25K
- □ < US$50K
- □ < US$100K
- □ > US$100K

Sales Channel
- □ Direct Only
- □ VAR Only
- ■ Direct and VAR
- □ Big Six / Outsource Firm

Direct Sales & Support Offices
- ■ In USA/Canada
- ■ In UK/Europe
- ■ In Pacific Rim/Australasia
- ■ In Central/South America

Skylight Systems

Address Data
Skylight Systems
141 Greenwood Ave.
Wyncote PA 19095

Contact Data
Tel: (215) 576-1001
Fax: (215) 576-1527
Web: http://www.skylightsystems.com

Facts and Figures
Founded: 1988
Country: USA
WW User Base: N/A
'95 Revenues ($M): N/A

Relational Financial Systems (RFS)

Financials
- ■ General Ledger
- ■ Financial Reporting
- ■ Consolidations
- ■ Budgeting
- ☐ Project Management
- ■ Fixed Assets

Supply Chain
- ■ Accounts Payable
- ■ Purchasing
- ■ Accounts Receivable
- ■ Order Entry   ■ Billing
- ■ Inventory   ☐ Bill of Materials
- ■ EDI

Other
- ☐ Manufacturing
- ☐ Human Resources
- ☐ Payroll & Benefits
- ■ Multi Currency
- ☐ Multi Lingual
- ■ Allocations

C/S PLATFORM SUPPORT

Server Operating System
- ☐ UNIX - IBM
- ■ UNIX - HP
- ■ UNIX - SUN
- ■ Microsoft NT Server
- ■ Novell Netware
- ☐ IBM AS/400 (OS/400)
- ☐ IBM Mainframe

RDBMS Engine
- ■ Oracle 7
- ■ Informix
- ■ Sybase 11
- ■ Microsoft SQL Server 6.x
- ☐ Centura SQLBase
- ☐ BTI Scaleable SQL/Btrieve
- ☐ IBM DB2 (any platform)

Client GUI
- ■ Windows 95/NT (32 bit)
- ■ Windows 3.x (16 bit)
- ☐ Apple MACOS
- ☐ IBM OS/2 Warp (32 bit)
- ■ OSF/Motif
- ■ Sun OpenLook
- ■ Character

TECHNOLOGY SUPPORT

Development Technology
- ■ C   ☐ C++
- ☐ COBOL   ☐ RPG
- ☐ SmallTalk   ☐ Other OO
- ☐ Progress   ☐ NewEra
- ☐ PowerBuilder   ☐ SQL Windows
- ☐ MS Access   ☐ Visual Basic
- ■ Other 4GL: Passport

Application Technology
- ☐ E mail enabled
- ☐ Workflow enabled
- ■ Document/image enabled
- ☐ Telephony enabled
- ☐ OLAP connectivity or module
- ■ User accessible Data Dictionary
- ■ Internet/Intranet access enabled

Microsoft Technology
- ■ MAPI Compliant
- ☐ MFC Used
- ■ ODBC Compliant
- ☐ OLE 2/ActiveX Complaint
- ☐ Designed for Windows 95 Logo
- ☐ Office Compatible Logo
- ☐ BackOffice Powered Logo

MARKETING INFORMATION

Start Price Per Module
- ■ < US$25K
- ☐ < US$50K
- ☐ < US$100K
- ☐ > US$100K

Sales Channel
- ☐ Direct Only
- ☐ VAR Only
- ■ Direct and VAR
- ☐ Big Six / Outsource Firm

Direct Sales & Support Offices
- ■ In USA/Canada
- ■ In UK/Europe
- ☐ In Pacific Rim/Australasia
- ☐ In Central/South America

Software 2000 Inc.

Address Data	**Contact Data**	**Facts and Figures**
Software 2000 Inc.	Tel: (508) 778-2000	Founded: 1981
25 Communications Way	Fax: (508) 778-5420	Country: USA
Drawer 6000	Web: http://www.s2k.com	Status: Public
Hyannis MA 02601		WW User Base: 1300
		'96 Revenues ($M): 72

Infinium Financial Management

Financials	**Supply Chain**	**Other**
■ General Ledger	■ Accounts Payable	■ Manufacturing
■ Financial Reporting	■ Purchasing	■ Human Resources
■ Consolidations	■ Accounts Receivable	■ Payroll & Benefits
■ Budgeting	■ Order Entry ■ Billing	■ Multi Currency
■ Project Management	■ Inventory ■ Bill of Materials	■ Multi Lingual
■ Fixed Assets	■ EDI	■ Allocations

C/S PLATFORM SUPPORT

Server Operating System	**RDBMS Engine**	**Client GUI**
☐ UNIX - IBM	☐ Oracle 7	■ Windows 95/NT (32 bit)
☐ UNIX - HP	☐ Informix	■ Windows 3.x (16 bit)
☐ UNIX - SUN	☐ Sybase 11	☐ Apple MACOS
☐ Microsoft NT Server	☐ Microsoft SQL Server 6.x	■ IBM OS/2 Warp (32 bit)
☐ Novell Netware	☐ Centura SQLBase	☐ OSF/Motif
■ IBM AS/400 (OS/400)	☐ BTI Scaleable SQL/Btrieve	☐ Sun OpenLook
☐ IBM Mainframe	■ IBM DB2 (any platform)	■ Character

TECHNOLOGY SUPPORT

Development Technology		**Application Technology**	**Microsoft Technology**
■ C	■ C++	■ E mail enabled	☐ MAPI Compliant
☐ COBOL	■ RPG	■ Workflow enabled	☐ MFC Used
■ SmallTalk	☐ Other OO	■ Document/image enabled	■ ODBC Compliant
☐ Progress	☐ NewEra	■ Telephony enabled	☐ OLE 2/ActiveX Complaint
☐ PowerBuilder	☐ SQL Windows	■ OLAP connectivity or module	☐ Designed for Windows 95 Logo
☐ MS Access	☐ Visual Basic	■ User accessible Data Dictionary	☐ Office Compatible Logo
☐ Other 4GL		■ Internet/Intranet access enabled	☐ BackOffice Powered Logo

MARKETING INFORMATION

Start Price Per Module	**Sales Channel**	**Direct Sales & Support Offices**
☐ < US$25K	■ Direct Only	■ In USA/Canada
■ < US$50K	☐ VAR Only	■ In UK/Europe
■ < US$100K	☐ Direct and VAR	■ In Pacific Rim/Australasia
■ > US$100K	■ Big Six / Outsource Firm	■ In Central/South America

Solomon Software

Address Data	**Contact Data**	**Facts and Figures**
Solomon Software	Tel: (419) 424-0422	Founded: 1980
200 East Hardin St.	Fax: (419) 424-3400	Country: USA
Findlay OH 45840	Web: http://www.solomon.com	Status: Private
		WW User Base: 45000+
		'95 Revenues ($M): N/A

Solomon IV for Windows

Financials	**Supply Chain**	**Other**
■ General Ledger	■ Accounts Payable	▢ Manufacturing
■ Financial Reporting	■ Purchasing	▢ Human Resources
■ Consolidations	■ Accounts Receivable	■ Payroll & Benefits
■ Budgeting	■ Order Entry ■ Billing	■ Multi Currency
■ Project Management	■ Inventory ■ Bill of Materials	▢ Multi Lingual
▢ Fixed Assets	▢ EDI	■ Allocations

C/S PLATFORM SUPPORT

Server Operating System	**RDBMS Engine**	**Client GUI**
▢ UNIX - IBM	▢ Oracle 7	■ Windows 95/NT (32 bit)
▢ UNIX - HP	▢ Informix	■ Windows 3.x (16 bit)
▢ UNIX - SUN	▢ Sybase 11	▢ Apple MACOS
■ Microsoft NT Server	■ Microsoft SQL Server 6.x	▢ IBM OS/2 Warp (32 bit)
■ Novell Netware	▢ Centura SQLBase	▢ OSF/Motif
▢ IBM AS/400 (OS/400)	■ BTI Scaleable SQL/Btrieve	▢ Sun OpenLook
▢ IBM Mainframe	▢ IBM DB2 (any platform)	▢ Character

TECHNOLOGY SUPPORT

Development Technology		**Application Technology**	**Microsoft Technology**
▢ C	▢ C++	▢ E mail enabled	▢ MAPI Compliant
▢ COBOL	▢ RPG	▢ Workflow enabled	▢ MFC Used
▢ SmallTalk	▢ Other OO	▢ Document/image enabled	■ ODBC Compliant
▢ Progress	▢ NewEra	▢ Telephony enabled	▢ OLE 2/ActiveX Complaint
▢ PowerBuilder	▢ SQL Windows	▢ OLAP connectivity or module	▢ Designed for Windows 95 Logo
▢ MS Access	■ Visual Basic	▢ User accessible Data Dictionary	▢ Office Compatible Logo
▢ Other 4GL		▢ Internet/Intranet access enabled	■ BackOffice Powered Logo

MARKETING INFORMATION

Start Price Per Module	**Sales Channel**	**Direct Sales & Support Offices**
■ < US$25K	▢ Direct Only	■ In USA/Canada
▢ < US$50K	■ VAR Only	■ In UK/Europe
▢ < US$100K	▢ Direct and VAR	■ In Pacific Rim/Australasia
▢ > US$100K	▢ Big Six / Outsource Firm	■ In Central/South America

SPFC

Address Data	Contact Data	Facts and Figures
SPFC	Tel: (800) 676-7732	Founded: 1990
P.O. Box 163	Fax: (201) 765-0791	Country: USA
Madison NJ 07940	Web: http://ourworld.compuserve.	Status: Private
	com/homepages/sqlaccounting	WW User Base: 80
		'95 Revenues ($M): N/A

SQL Accounting for Windows

Financials
- ■ General Ledger
- ■ Financial Reporting
- □ Consolidations
- □ Budgeting
- ■ Project Management
- ■ Fixed Assets

Supply Chain
- ■ Accounts Payable
- ■ Purchasing
- ■ Accounts Receivable
- ■ Order Entry   ■ Billing
- ■ Inventory   ■ Bill of Materials
- □ EDI

Other
- ■ Manufacturing
- □ Human Resources
- ■ Payroll & Benefits
- ■ Multi Currency
- □ Multi Lingual
- □ Allocations

C/S PLATFORM SUPPORT

Server Operating System
- ■ UNIX - IBM
- ■ UNIX - HP
- ■ UNIX - SUN
- ■ Microsoft NT Server
- ■ Novell Netware
- □ IBM AS/400 (OS/400)
- □ IBM Mainframe

RDBMS Engine
- ■ Oracle 7
- □ Informix
- ■ Sybase 11
- ■ Microsoft SQL Server 6.x
- ■ Centura SQLBase
- □ BTI Scaleable SQL/Btrieve
- □ IBM DB2 (any platform)

Client GUI
- ■ Windows 95/NT (32 bit)
- ■ Windows 3.x (16 bit)
- □ Apple MACOS
- □ IBM OS/2 Warp (32 bit)
- □ OSF/Motif
- □ Sun OpenLook
- □ Character

TECHNOLOGY SUPPORT

Development Technology
- □ C   □ C++
- □ COBOL   □ RPG
- □ SmallTalk   □ Other OO
- □ Progress   □ NewEra
- □ PowerBuilder   ■ SQL Windows
- □ MS Access   □ Visual Basic
- □ Other 4GL: Triton

Application Technology
- □ E mail enabled
- □ Workflow enabled
- □ Document/image enabled
- □ Telephony enabled
- □ OLAP connectivity or module
- ■ User accessible Data Dictionary
- □ Internet/Intranet access enabled

Microsoft Technology
- □ MAPI Compliant
- □ MFC Used
- □ ODBC Compliant
- □ OLE 2/ActiveX Complaint
- □ Designed for Windows 95 Logo
- □ Office Compatible Logo
- □ BackOffice Powered Logo

MARKETING INFORMATION

Start Price Per Module
- ■ < US$25K
- □ < US$50K
- □ < US$100K
- □ > US$100K

Sales Channel
- □ Direct Only
- □ VAR Only
- ■ Direct and VAR
- □ Big Six / Outsource Firm

Direct Sales & Support Offices
- ■ In USA/Canada
- □ In UK/Europe
- □ In Pacific Rim/Australasia
- □ In Central/South America

State of the Art Inc.

Address Data	**Contact Data**	**Facts and Figures**
State of the Art Inc.	Tel: (714) 753-1222	Founded: 1981
56 Technology	Fax: (714) 453-1650	Country: USA
Irvine CA 92718	Web: http://www.stateoftheart.com	Status: Public
		WW User Base: N/A
		'95 Revenues ($M): N/A

Acuity Financials

Financials
- ■ General Ledger
- ■ Financial Reporting
- □ Consolidations
- ■ Budgeting
- □ Project Management
- ▤ Fixed Assets

Supply Chain
- ■ Accounts Payable
- □ Purchasing
- ■ Accounts Receivable
- □ Order Entry □ Billing
- □ Inventory □ Bill of Materials
- □ EDI

Other
- □ Manufacturing
- ▤ Human Resources
- ▤ Payroll & Benefits
- ■ Multi Currency
- □ Multi Lingual
- ■ Allocations

C/S PLATFORM SUPPORT

Server Operating System
- □ UNIX - IBM
- □ UNIX - HP
- □ UNIX - SUN
- ■ Microsoft NT Server
- □ Novell Netware
- □ IBM AS/400 (OS/400)
- □ IBM Mainframe

RDBMS Engine
- □ Oracle 7
- □ Informix
- □ Sybase 11
- ■ Microsoft SQL Server 6.x
- □ Centura SQLBase
- □ BTI Scaleable SQL/Btrieve
- □ IBM DB2 (any platform)

Client GUI
- ■ Windows 95/NT (32 bit)
- □ Windows 3.x (16 bit)
- □ Apple MACOS
- □ IBM OS/2 Warp (32 bit)
- □ OSF/Motif
- □ Sun OpenLook
- □ Character

TECHNOLOGY SUPPORT

Development Technology
- □ C ■ C++
- □ COBOL □ RPG
- □ SmallTalk □ Other OO
- □ Progress □ NewEra
- □ PowerBuilder □ SQL Windows
- □ MS Access ■ Visual Basic
- □ Other 4GL

Application Technology
- ■ E mail enabled
- ■ Workflow enabled
- ▤ Document/image enabled
- □ Telephony enabled
- □ OLAP connectivity or module
- ▤ User accessible Data Dictionary
- □ Internet/Intranet access enabled

Microsoft Technology
- ■ MAPI Compliant
- ■ MFC Used
- ■ ODBC Compliant
- ■ OLE 2/ActiveX Complaint
- □ Designed for Windows 95 Logo
- □ Office Compatible Logo
- □ BackOffice Powered Logo

MARKETING INFORMATION

Start Price Per Module
- ■ < US$25K
- □ < US$50K
- □ < US$100K
- □ > US$100K

Sales Channel
- □ Direct Only
- ■ VAR Only
- □ Direct and VAR
- □ Big Six / Outsource Firm

Direct Sales & Support Offices
- ■ In USA/Canada
- □ In UK/Europe
- □ In Pacific Rim/Australasia
- □ In Central/South America

SQL Financials International Inc.

Address Data	**Contact Data**	**Facts and Figures**
SQL Financials Intl. Inc.	Tel: (770) 390-3900	Founded: 1991
2 Ravinia Drive, Ste. 1000	Fax: (770) 390-3999	Country: USA
Atlanta GA 30346	Web: http://www.sqlfinancials.com	Status: Private
		WW User Base: 200+
		'95 Revenues ($M): N/A

SQL Financials

Financials
- ■ General Ledger
- ■ Financial Reporting
- ■ Consolidations
- ■ Budgeting
- ☐ Project Management
- ■ Fixed Assets

Supply Chain
- ■ Accounts Payable
- ■ Purchasing
- ■ Accounts Receivable
- ☐ Order Entry ■ Billing
- ☐ Inventory ☐ Bill of Materials
- ■ EDI

Other
- ☐ Manufacturing
- ■ Human Resources
- ■ Payroll & Benefits
- ■ Multi Currency
- ☐ Multi Lingual
- ■ Allocations

C/S PLATFORM SUPPORT

Server Operating System
- ■ UNIX - IBM
- ■ UNIX - HP
- ■ UNIX - SUN
- ■ Microsoft NT Server
- ■ Novell Netware
- ☐ IBM AS/400 (OS/400)
- ☐ IBM Mainframe

RDBMS Engine
- ■ Oracle 7
- ☐ Informix
- ■ Sybase 11
- ■ Microsoft SQL Server 6.x
- ☐ Centura SQLBase
- ☐ BTI Scaleable SQL/Btrieve
- ☐ IBM DB2 (any platform)

Client GUI
- ■ Windows 95/NT (32 bit)
- ■ Windows 3.x (16 bit)
- ☐ Apple MACOS
- ☐ IBM OS/2 Warp (32 bit)
- ☐ OSF/Motif
- ☐ Sun OpenLook
- ☐ Character

TECHNOLOGY SUPPORT

Development Technology
- ☐ C ■ C++
- ☐ COBOL ☐ RPG
- ☐ SmallTalk ☐ Other OO
- ☐ Progress ☐ NewEra
- ☐ PowerBuilder ■ SQL Windows
- ☐ MS Access ☐ Visual Basic
- ☐ Other 4GL:

Application Technology
- ■ E mail enabled
- ■ Workflow enabled
- ■ Document/image enabled
- ☐ Telephony enabled
- ■ OLAP connectivity or module
- ☐ User accessible Data Dictionary
- ☐ Internet/Intranet access enabled

Microsoft Technology
- ■ MAPI Compliant
- ■ MFC Used
- ■ ODBC Compliant
- ■ OLE 2/ActiveX Complaint
- ■ Designed for Windows 95 Logo
- ☐ Office Compatible Logo
- ☐ BackOffice Powered Logo

MARKETING INFORMATION

Start Price Per Module
- ☐ < US$25K
- ☐ < US$50K
- ■ < US$100K
- ☐ > US$100K

Sales Channel
- ■ Direct Only
- ☐ VAR Only
- ☐ Direct and VAR
- ☐ Big Six / Outsource Firm

Direct Sales & Support Offices
- ■ In USA/Canada
- ☐ In UK/Europe
- ☐ In Pacific Rim/Australasia
- ☐ In Central/South America

Synon Corporation

Address Data
Synon Corporation
1100 Larkspur Landing Circle
Larkspur CA 94939

Contact Data
Tel: (415) 451-5000
Fax: (415) 481-4239
Web: http://www.synon.com

Facts and Figures
Founded: 1984
Country: USA
Status: Private
WW User Base: 8000
'95 Revenues ($M): 80

Synon Financials

Financials
- ■ General Ledger
- ■ Financial Reporting
- ■ Consolidations
- ■ Budgeting
- □ Project Management
- ■ Fixed Assets

Supply Chain
- ■ Accounts Payable
- □ Purchasing
- ■ Accounts Receivable
- □ Order Entry □ Billing
- □ Inventory □ Bill of Materials
- □ EDI

Other
- □ Manufacturing
- □ Human Resources
- □ Payroll & Benefits
- ■ Multi Currency
- ■ Multi Lingual
- ■ Allocations

C/S PLATFORM SUPPORT

Server Operating System
- □ UNIX - IBM
- ■ UNIX - HP
- □ UNIX - SUN
- ■ Microsoft NT Server
- □ Novell Netware
- ■ IBM AS/400 (OS/400)
- □ IBM Mainframe

RDBMS Engine
- ■ Oracle 7
- □ Informix
- □ Sybase 11
- ■ Microsoft SQL Server 6.x
- □ Centura SQLBase
- □ BTI Scaleable SQL/Btrieve
- □ IBM DB2 (any platform)

Client GUI
- ■ Windows 95/NT (32 bit)
- ■ Windows 3.x (16 bit)
- □ Apple MACOS
- □ IBM OS/2 Warp (32 bit)
- □ OSF/Motif
- □ Sun OpenLook
- ■ Character

TECHNOLOGY SUPPORT

Development Technology
- □ C ■ C++
- ■ COBOL ■ RPG
- □ SmallTalk □ Other OO
- □ Progress □ NewEra
- □ PowerBuilder □ SQL Windows
- □ MS Access □ Visual Basic
- ■ Other 4GL: Synon 2E & Obsydian

Application Technology
- ■ E mail enabled
- □ Workflow enabled
- □ Document/image enabled
- □ Telephony enabled
- ■ OLAP connectivity or module
- ■ User accessible Data Dictionary
- ■ Internet/Intranet access enabled

Microsoft Technology
- ■ MAPI Compliant
- ■ MFC Used
- ■ ODBC Compliant
- □ OLE 2/ActiveX Complaint
- □ Designed for Windows 95 Logo
- □ Office Compatible Logo
- ■ BackOffice Powered Logo

MARKETING INFORMATION

Start Price Per Module
- ■ < US$25K
- □ < US$50K
- □ < US$100K
- □ > US$100K

Sales Channel
- □ Direct Only
- □ VAR Only
- ■ Direct and VAR
- □ Big Six / Outsource Firm

Direct Sales & Support Offices
- ■ In USA/Canada
- ■ In UK/Europe
- ■ In Pacific Rim/Australasia
- ■ In Central/South America

Syspro Group

Address Data	Contact Data	Facts and Figures
	Tel: (800) 369-8649	Founded: 1977
	Fax: (714) 437-1000	Country: USA/SA
	Web: http://www.sysprousa.com	Status:
		WWW User Base: 5000
		'95 Revenues ($M): 25

Impact Encore/Award

Financials
- ■ General Ledger
- ■ Financial Reporting
- ■ Consolidations
- ■ Budgeting
- ■ Project Management
- ■ Fixed Assets

Supply Chain
- ■ Accounts Payable
- ■ Purchasing 2Q/97)
- ■ Accounts Receivable
- ■ Order Entry   ■ Billing
- ■ Inventory   ■ Bill of Materials
- ■ EDI

Other
- ■ Manufacturing
- ■ Human Resources
- ■ Payroll & Benefits
- ■ Multi Currency
- ☐ Multi Lingual
- ■ Allocations

C/S PLATFORM SUPPORT

Server Operating System
- ■ UNIX - IBM
- ■ UNIX - HP
- ■ UNIX - SUN
- ■ Microsoft NT Server
- ■ Novell Netware
- ☐ IBM AS/400 (OS/400)
- ☐ IBM Mainframe

RDBMS Engine
- ☐ Oracle 7
- ☐ Informix
- ☐ Sybase 11
- ☐ Microsoft SQL Server 6.x
- ☐ Centura SQLBase
- ☐ BTI Scaleable SQL/Btrieve
- ☐ IBM DB2 (any platform)

Client GUI
- ■ Windows 95/NT (32 bit)
- ■ Windows 3.x (16 bit)
- ☐ Apple MACOS
- ☐ IBM OS/2 Warp (32 bit)
- ☐ OSF/Motif
- ☐ Sun OpenLook
- ■ Character

TECHNOLOGY SUPPORT

Development Technology
- ☐ C
- ■ COBOL
- ☐ SmallTalk
- ☐ Progress
- ☐ PowerBuilder
- ☐ MS Access
- ☐ Other 4GL
- ☐ C++
- ☐ RPG
- ☐ Other OO
- ☐ NewEra
- ☐ SQL Windows
- ☐ Visual Basic

Application Technology
- ■ E mail enabled
- ☐ Workflow enabled
- ■ Document/image enabled
- ☐ Telephony enabled
- ■ OLAP connectivity or module
- ■ User accessible Data Dictionary
- ■ Internet/Intranet access enabled

Microsoft Technology
- ☐ MAPI Compliant
- ☐ MFC Used
- ■ ODBC Compliant
- ■ OLE 2/ActiveX Complaint
- ■ Designed for Windows 95 Logo
- ■ Office Compatible Logo
- ■ BackOffice Powered Logo

MARKETING INFORMATION

Start Price Per Module
- ■ < US$25K
- ☐ < US$50K
- ☐ < US$100K
- ☐ > US$100K

Sales Channel
- ☐ Direct Only
- ■ VAR Only
- ☐ Direct and VAR
- ☐ Big Six / Outsource Firm

Direct Sales & Support Offices
- ■ In USA/Canada
- ■ In UK/Europe
- ■ In Pacific Rim/Australasia
- ☐ In Central/South America

System Software Associates Inc.

Address Data
System Software Associates Inc.
500 West Madison
Chicago IL 60661

Contact Data
Tel: (312) 258-6000
Fax: (312) 474-7500
Web: http://www.ssax.com

Facts and Figures
Founded: 1981
Country: USA
WW User Base: 8000
'95 Revenues ($M): 394.4

BPCS Client/Server

Financials
- ■ General Ledger
- ■ Financial Reporting
- ■ Consolidations
- ■ Budgeting
- ■ Project Management
- ■ Fixed Assets

Supply Chain
- ■ Accounts Payable
- ■ Purchasing
- ■ Accounts Receivable
- ■ Order Entry ■ Billing
- ■ Inventory ■ Bill of Materials
- ■ EDI

Other
- ■ Manufacturing
- ☐ Human Resources
- ☐ Payroll & Benefits
- ■ Multi Currency
- ■ Multi Lingual
- ■ Allocations

C/S PLATFORM SUPPORT

Server Operating System
- ■ UNIX - IBM
- ■ UNIX - HP
- ☐ UNIX - SUN
- ☐ Microsoft NT Server
- ☐ Novell Netware
- ■ IBM AS/400 (OS/400)
- ☐ IBM Mainframe

RDBMS Engine
- ■ Oracle 7
- ■ Informix
- ☐ Sybase 11
- ■ Microsoft SQL Server 6.x
- ☐ Centura SQLBase
- ☐ BTI Scaleable SQL/Btrieve
- ■ IBM DB2 (any platform)

Client GUI
- ■ Windows 95/NT (32 bit)
- ■ Windows 3.x (16 bit)
- ☐ Apple MACOS
- ☐ IBM OS/2 Warp (32 bit)
- ☐ OSF/Motif
- ☐ Sun OpenLook
- ☐ Character

TECHNOLOGY SUPPORT

Development Technology
- ■ C ■ C++
- ☐ COBOL ■ RPG
- ☐ SmallTalk ■ Other OO
- ☐ Progress ☐ NewEra
- ☐ PowerBuilder ☐ SQL Windows
- ☐ MS Access ☐ Visual Basic
- ☐ Other 4GL

Application Technology
- ■ E mail enabled
- ■ Workflow enabled
- ■ Document/image enabled
- ■ Telephony enabled
- ■ OLAP connectivity or module
- ■ User accessible Data Dictionary
- ■ Internet/Intranet access enabled

Microsoft Technology
- ■ MAPI Compliant
- ☐ MFC Used
- ■ ODBC Compliant
- ■ OLE 2/ActiveX Complaint
- ☐ Designed for Windows 95 Logo
- ☐ Office Compatible Logo
- ☐ BackOffice Powered Logo

MARKETING INFORMATION

Start Price Per Module
- ■ < US$25K
- ■ < US$50K
- ■ < US$100K
- ■ > US$100K

Sales Channel
- ■ Direct Only
- ☐ VAR Only
- ☐ Direct and VAR
- ☐ Big Six / Outsource Firm

Direct Sales & Support Offices
- ■ In USA/Canada
- ■ In UK/Europe
- ■ In Pacific Rim/Australasia
- ■ In Central/South America

Systems Union Inc.

Address Data	**Contact Data**	**Facts and Figures**
Systems Union Inc.	Tel: (914) 948-7770	Founded: 1981
10 Bank St.	Fax: (914) 948-7399	Country: UK
White Plains NY 10606	Web: http://www.systemsunion.com	Status: Private
		WW User Base: 13000
		'95 Revenues ($M): 59.3

SunSystems

Financials
- ■ General Ledger
- ■ Financial Reporting
- ■ Consolidations
- ■ Budgeting
- ■ Project Management
- ■ Fixed Assets

Supply Chain
- ■ Accounts Payable
- ■ Purchasing
- ■ Accounts Receivable
- ■ Order Entry   ■ Billing
- ■ Inventory   □ Bill of Materials
- □ EDI

Other
- □ Manufacturing
- □ Human Resources
- □ Payroll & Benefits
- ■ Multi Currency
- ■ Multi Lingual
- ■ Allocations

C/S PLATFORM SUPPORT

Server Operating System
- ■ UNIX - IBM
- ■ UNIX - HP
- ■ UNIX - SUN
- ■ Microsoft NT Server
- ■ Novell Netware
- □ IBM AS/400 (OS/400)
- □ IBM Mainframe

RDBMS Engine
- ■ Oracle 7
- □ Informix
- ■ Sybase 11
- ■ Microsoft SQL Server 6.x
- □ Centura SQLBase
- ■ BTI Scaleable SQL/Btrieve
- □ IBM DB2 (any platform)

Client GUI
- ■ Windows 95/NT (32 bit)
- ■ Windows 3.x (16 bit)
- □ Apple MACOS
- □ IBM OS/2 Warp (32 bit)
- □ OSF/Motif
- □ Sun OpenLook
- ■ Character

TECHNOLOGY SUPPORT

Development Technology
- □ C □ C++
- ■ COBOL □ RPG
- □ SmallTalk □ Other OO
- □ Progress □ NewEra
- □ PowerBuilder □ SQL Windows
- □ MS Access □ Visual Basic
- □ Other 4GL

Application Technology
- □ E mail enabled
- □ Workflow enabled
- □ Document/image enabled
- □ Telephony enabled
- ■ OLAP connectivity or module
- □ User accessible Data Dictionary
- □ Internet/Intranet access enabled

Microsoft Technology
- □ MAPI Compliant
- □ MFC Used
- □ ODBC Compliant
- □ OLE 2/ActiveX Complaint
- □ Designed for Windows 95 Logo
- □ Office Compatible Logo
- □ BackOffice Powered Logo

MARKETING INFORMATION

Start Price Per Module
- ■ < US$25K
- □ < US$50K
- □ < US$100K
- □ > US$100K

Sales Channel
- □ Direct Only
- □ VAR Only
- ■ Direct and VAR
- □ Big Six / Outsource Firm

Direct Sales & Support Offices
- ■ In USA/Canada
- ■ In UK/Europe
- ■ In Pacific Rim/Australasia
- ■ In Central/South America

Timeline Inc.

Address Data	**Contact Data**	**Facts and Figures**
Timeline Inc.	Tel: (206) 822-3140	Founded: 1977
3055 112th Ave. NE, Ste. 106	Fax: (206) 822-1120	Country: USA
Bellevue WA 98004	Web: http://www.timeline.com	Status: Public
		WW User Base: 300
		'95 Revenues ($M): 6.5

MetaView

Financials
- ☐ General Ledger
- ■ Financial Reporting
- ■ Consolidations
- ■ Budgeting
- ☐ Project Management
- ☐ Fixed Assets

Supply Chain
- ☐ Accounts Payable
- ☐ Purchasing
- ☐ Accounts Receivable
- ☐ Order Entry   ☐ Billing
- ☐ Inventory   ☐ Bill of Materials
- ☐ EDI

Other
- ☐ Manufacturing
- ☐ Human Resources
- ☐ Payroll & Benefits
- ■ Multi Currency
- ☐ Multi Lingual
- ■ Allocations

C/S PLATFORM SUPPORT

Server Operating System
- ☐ UNIX - IBM
- ☐ UNIX - HP
- ☐ UNIX - SUN
- ■ Microsoft NT Server
- ☐ Novell Netware
- ☐ IBM AS/400 (OS/400)
- ☐ IBM Mainframe

RDBMS Engine
- ☐ Oracle 7
- ☐ Informix
- ☐ Sybase 11
- ■ Microsoft SQL Server 6.x
- ☐ Centura SQLBase
- ☐ BTI Scaleable SQL/Btrieve
- ☐ IBM DB2 (any platform)

Client GUI
- ■ Windows 95/NT (32 bit)
- ■ Windows 3.x (16 bit)
- ☐ Apple MACOS
- ☐ IBM OS/2 Warp (32 bit)
- ☐ OSF/Motif
- ☐ Sun OpenLook
- ☐ Character

TECHNOLOGY SUPPORT

Development Technology
- ☐ C   ■ C++
- ☐ COBOL   ☐ RPG
- ☐ SmallTalk ☐ Other OO
- ☐ Progress   ☐ NewEra
- ☐ PowerBuilder ☐ SQL Windows
- ☐ MS Access ■ Visual Basic
- ☐ Other 4GL

Application Technology
- ☐ E mail enabled
- ☐ Workflow enabled
- ☐ Document/image enabled
- ☐ Telephony enabled
- ■ OLAP connectivity or module
- ■ User accessible Data Dictionary
- ■ Internet/Intranet access enabled

Microsoft Technology
- ■ MAPI Compliant
- ■ MFC Used
- ☐ ODBC Compliant
- ☐ OLE 2/ActiveX Complaint
- ■ Designed for Windows 95 Logo
- ■ Office Compatible Logo
- ☐ BackOffice Powered Logo

MARKETING INFORMATION

Start Price Per Module
- ☐ Please see a sales representative

Sales Channel
- ☐ Direct Only
- ☐ VAR Only
- ■ Direct and VAR
- ■ Big Six / Outsource Firm

Direct Sales & Support Offices
- ■ In USA/Canada
- ■ In UK/Europe
- ☐ In Pacific Rim/Australasia
- ☐ In Central/South America

USL Systems

Address Data
USL Systems
501 Church St. NE
Vienna VA 22180

Contact Data
Tel: (703) 242-0402
Fax: (703) 242-0402
Web: http://www.uslsystems.com

Facts and Figures
Founded: 1988
Country: USA
Status: Public
WW User Base: 40
'95 Revenues ($M): N/A

USL Financials

Financials
- ■ General Ledger
- ■ Financial Reporting
- ■ Consolidations
- ■ Budgeting
- ■ Project Management
- ▣ Fixed Assets

Supply Chain
- ■ Accounts Payable
- ■ Purchasing
- ■ Accounts Receivable
- ■ Order Entry   ■ Billing
- ■ Inventory   □ Bill of Materials
- □ EDI

Other
- □ Manufacturing
- ▣ Human Resources
- ▣ Payroll & Benefits
- □ Multi Currency
- □ Multi Lingual
- ■ Allocations

C/S PLATFORM SUPPORT

Server Operating System
- □ UNIX - IBM
- ■ UNIX - HP
- ■ UNIX - SUN
- ■ Microsoft NT Server
- ■ Novell Netware
- □ IBM AS/400 (OS/400)
- □ IBM Mainframe

RDBMS Engine
- □ Oracle 7
- □ Informix
- ■ Sybase 11
- ■ Microsoft SQL Server 6.x
- □ Centura SQLBase
- □ BTI Scaleable SQL/Btrieve
- □ IBM DB2 (any platform)

Client GUI
- ■ Windows 95/NT (32 bit)
- ■ Windows 3.x (16 bit)
- □ Apple MACOS
- ■ IBM OS/2 Warp (32 bit)
- □ OSF/Motif
- □ Sun OpenLook
- □ Character

TECHNOLOGY SUPPORT

Development Technology
- ■ C   □ C++
- □ COBOL   □ RPG
- □ SmallTalk   □ Other OO
- □ Progress   □ NewEra
- □ PowerBuilder □ SQL Windows
- ■ MS Access   □ Visual Basic
- □ Other 4GL

Application Technology
- ■ E mail enabled
- ■ Workflow enabled
- □ Document/image enabled
- □ Telephony enabled
- ■ OLAP connectivity or module
- □ User accessible Data Dictionary
- □ Internet/Intranet access enabled

Microsoft Technology
- ■ MAPI Compliant
- □ MFC Used
- ■ ODBC Compliant
- ■ OLE 2/ActiveX Complaint
- □ Designed for Windows 95 Logo
- □ Office Compatible Logo
- □ BackOffice Powered Logo

MARKETING INFORMATION

Start Price Per Module
- ■ < US$25K
- □ < US$50K
- □ < US$100K
- □ > US$100K

Sales Channel
- □ Direct Only
- □ VAR Only
- ■ Direct and VAR
- □ Big Six / Outsource Firm

Direct Sales & Support Offices
- ■ In USA/Canada
- □ In UK/Europe
- □ In Pacific Rim/Australasia
- □ In Central/South America

Walker Interactive Systems

Address Data	**Contact Data**	**Facts and Figures**
Walker Interactive Systems	Tel: (415) 495-8811	Founded: 1969
Marathon Plaza Three North	Fax: (415) 543-6338	Country: USA
303 Second Street	Web: http://www.walker/com	Status: Public
San Francisco CA 94107		WW User Base: 950
		'95 Revenues ($M): 60

Tamaris C/S and Financial Management and Reporting

Financials
- ■ General Ledger
- ■ Financial Reporting
- ■ Consolidations
- ■ Budgeting
- ■ Project Management
- ■ Fixed Assets

Supply Chain
- ■ Accounts Payable
- ■ Purchasing
- ■ Accounts Receivable
- □ Order Entry □ Billing
- ■ Inventory □ Bill of Materials
- 🖫 EDI

Other
- □ Manufacturing
- □ Human Resources
- □ Payroll & Benefits
- ■ Multi Currency
- □ Multi Lingual
- □ Allocations

C/S PLATFORM SUPPORT

Server Operating System
- □ UNIX - IBM
- □ UNIX - HP
- □ UNIX - SUN
- ■ Microsoft NT Server
- □ Novell Netware
- □ IBM AS/400 (OS/400)
- ■ IBM Mainframe

RDBMS Engine
- □ Oracle 7
- □ Informix
- □ Sybase 11
- □ Microsoft SQL Server 6.x
- □ Centura SQLBase
- □ BTI Scaleable SQL/Btrieve
- ■ IBM DB2 (any platform)

Client GUI
- ■ Windows 95/NT (32 bit)
- ■ Windows 3.x (16 bit)
- □ Apple MACOS
- □ IBM OS/2 Warp (32 bit)
- □ OSF/Motif
- □ Sun OpenLook
- ■ Character

TECHNOLOGY SUPPORT

Development Technology
- ■ C
- ■ COBOL
- □ SmallTalk
- □ Progress
- □ PowerBuilder
- □ MS Access
- □ Other 4GL
- ■ C++
- □ RPG
- □ Other OO
- □ NewEra
- □ SQL Windows
- ■ Visual Basic

Application Technology
- 🖫 E mail enabled
- ■ Workflow enabled
- 🖫 Document/image enabled
- □ Telephony enabled
- ■ OLAP connectivity or module
- □ User accessible Data Dictionary
- □ Internet/Intranet access enabled

Microsoft Technology
- □ MAPI Compliant
- □ MFC Used
- ■ ODBC Compliant
- 🖫 OLE 2/ActiveX Complaint
- □ Designed for Windows 95 Logo
- □ Office Compatible Logo
- □ BackOffice Powered Logo

MARKETING INFORMATION

Start Price Per Module
- □ < US$25K
- □ < US$50K
- ■ < US$100K
- ■ > US$100K

Sales Channel
- ■ Direct Only
- □ VAR Only
- □ Direct and VAR
- □ Big Six / Outsource Firm

Direct Sales & Support Offices
- ■ In USA/Canada
- ■ In UK/Europe
- ■ In Pacific Rim/Australasia
- □ In Central/South America

Chapter 14

Vendor Cross-References

These cross-references are compiled from the 54 vendor profiles contained in Chapter 13. There are nine cross-references.

1. Financial modules cross-reference
2. Supply chain modules cross-reference
3. Other modules cross-reference
4. Server support cross-reference
5. RDBMS support cross-reference
6. GUI support cross-reference
7. Pricing cross-reference
8. Sales channels cross-reference
9. Sales and support cross-reference

Financial Modules Cross-Reference

	General Ledger	Financial Reporting	Consolidation	Budgeting	Project Control	Fixed Assets
AccountMate Software	■	■	■	■	□	□
Agresso Corp.	■	■	■	■	■	■
American Software	■	■	■	■	■	□
Apprise Software Inc.	■	■	■	■	■	■
Baan Company	■	■	■	■	■	■
BHR Software	■	■	■	■	■	□
Big Software	■	■	□	■	□	□
Carillon Financials Corporation	■	■	□	■	□	■
CODA Inc.	■	■	■	■	■	■
Computer Associates Int'l. Inc.	■	■	■	■	■	■
Computron Software Inc.	■	■	□	□	■	■
Concepts Dynamic, Inc.	■	■	■	■	■	■
Deltek Systems Inc.	■	■	■	■	■	■
Design Data Systems	■	■	■	■	■	■
FlexiInternational Software Inc.	■	■	■	■	■	■
Fourth Shift Corporation	■	■	■	■	□	□
Geac SmartStream	■	■	■	■	□	□
GeacVisionShift	■	■	■	■	□	□
Great Plains Software	■	■	■	■	□	■
Hyperion Software Corp.	■	■	■	■	■	■
IET, Inc.	■	■	□	■	■	■
JBA International	■	■	■	■	■	■
JD Edwards	■	■	■	■	■	■
Lawson Software	■	■	■	■	■	■
Macola Inc.	■	■	■	■	■	■
Maconomy NE Inc.	■	■	■	■	■	■
MTX International Inc.	■	■	■	■	□	■
Navison Software US Inc.	■	■	■	■	■	■
OpenPlus International Inc.	■	■	■	■	■	■
Open Systems Inc.	■	■	■	■	■	□
Orange Systems	■	■	■	■	■	■
PeopleSoft	■	■	■	■	■	■
Platinum Software Corporation	■	■	■	■	□	■
PowerCerv	■	■	■	□	■	■
QSP Inc.	■	■	□	■	□	■
Ramco Systems Corp.	■	■	■	■	□	■
Ross Systems Inc.	■	■	■	■	■	■
SAP America Inc.	■	■	■	■	■	■
SAS Institute Inc.	□	■	■	■	□	□
SBT Accounting Systems	■	■	■	■	■	■
Scala North America Inc.	■	■	■	■	■	■
Skylight Systems	■	■	■	■	□	■
Software 2000 Inc.	■	■	■	■	■	■
Solomon Software	■	■	■	■	■	■
SPFC	■	■	□	□	■	■
State of the Art Inc.	■	■	□	■	□	■
SQL Financials	■	■	■	■	□	■
Synon Corporation	■	■	■	■	□	■
Syspro Group	■	■	■	■	■	■
System Software Associates Inc.	■	■	■	■	■	■
Systems Union Inc.	■	■	■	■	■	■
Timeline Inc.	□	■	■	■	□	□
USL Systems	■	■	■	■	■	■
Walker Interactive Systems	■	■	■	■	■	■

Supply Chain Modules Cross-Reference

	Accounts Payable	Purchasing	Accounts Receivable	Order Entry	Billing	Inventory	BOM	EDI
AccountMate Software	■	■	■	■	■	■	■	□
Agresso Corp.	■	■	■	■	■	■	□	■
American Software	■	■	■	■	■	■	■	■
Apprise Software Inc.	■	■	■	■	■	■	□	■
Baan Company	■	■	■	■	■	■	■	■
BHR Software	■	■	■	■	■	■	■	■
Big Software	■	■	■	■	■	■	■	□
Carillon Financials Corporation	■	■	■	■	■	■	□	■
CODA Inc.	■	■	■	□	□	□	□	□
Computer Associates Int'l Inc.	■	■	■	□	□	■	□	■
Computron Software Inc.	■	■	■	□	□	■	□	■
Concepts Dynamic, Inc.	■	■	■	■	■	■	■	■
Deltek Systems Inc.	■	■	■	■	■	■	■	■
Design Data Systems	■	■	■	■	■	■	■	□
FlexiInternational Software Inc.	■	■	■	■	□	■	□	■
Fourth Shift Corporation	■	■	■	■	□	□	■	■
Geac SmartStream	■	■	■	■	■	■	■	■
GeacVisionShift	■	■	■	□	■	■	□	■
Great Plains Software	■	■	■	■	■	■	■	■
Hyperion Software Corp.	■	■	■	□	□	□	□	■
IET, Inc.	■	■	■	■	■	■	□	□
JBA International	■	■	■	■	■	■	■	■
JD Edwards	■	■	■	■	■	■	■	■
Lawson Software	■	■	■	■	■	■	■	■
Macola Inc.	■	■	■	■	□	■	■	■
Maconomy NE Inc.	■	■	■	■	■	■	■	■
MTX International Inc.	■	■	■	■	■	■	□	□
Navison Software US Inc.	■	■	■	■	■	■	■	■
OpenPlus International Inc.	■	■	■	■	■	■	■	■
Open Systems Inc.	■	■	■	■	■	■	■	□
Orange Systems	■	■	■	■	■	■	■	□
PeopleSoft	■	■	■	■	■	■	■	■
Platinum Software Corporation	■	■	■	■	■	■	□	■
PowerCerv	■	■	■	■	■	■	■	■
QSP Inc.	■	■	■	□	□	□	□	□
Ramco Systems Corp.	■	■	■	■	■	■	■	■
Ross Systems Inc.	■	■	■	■	■	■	□	■
SAP America Inc.	■	■	■	■	■	■	■	■
SAS Institute Inc.	□	□	□	□	□	□	□	□
SBT Accounting Systems	■	■	■	■	■	■	■	□
Scala North America Inc.	■	■	■	■	■	■	■	■
Skylight Systems	■	■	■	■	■	■	□	■
Software 2000 Inc.	■	■	■	■	■	■	■	■
Solomon Software	■	■	■	■	■	■	■	■
SPFC	■	■	■	■	■	■	■	□
State of the Art Inc.	■	□	■	□	□	□	□	■
SQL Financials	■	■	■	□	□	□	□	■
Synon Corporation	■	□	■	□	□	□	□	■
Syspro Group	■	■	■	■	■	■	■	■
System Software Associates Inc.	■	■	■	■	■	■	■	■
Systems Union Inc.	■	■	■	■	■	■	□	□
Timeline Inc.	□	□	□	□	□	□	□	□
USL Systems	■	■	■	■	■	■	□	□
Walker Interactive Systems	■	■	■	□	□	■	□	■

Other Modules Cross-Reference

	Manufacturing	Human Resources	Payroll/ Benefits	Multi Currency	Multi Lingual	Allocations
AccountMate Software	■	□	■	■	■	■
Agresso Corp.	□	■	■	■	■	■
American Software	■	□	□	■	■	■
Apprise Software Inc.	□	□	□	■	■	■
Baan Company	■	□	□	■	■	□
BHR Software	■	□	□	□	□	□
Big Software	■	□	■	□	□	□
Carillon Financials Corporation	□	□	□	■	■	■
CODA Inc.	□	□	□	■	■	■
Computer Associates Int'l Inc.	□	□	□	■	■	□
Computron Software Inc.	□	□	□	■	■	■
Concepts Dynamic, Inc.	■	■	■	■	■	■
Deltek Systems Inc.	■	■	■	■	□	■
Design Data Systems	□	□	■	■	□	■
FlexiInternational Software Inc.	□	□	□	■	■	■
Fourth Shift Corporation	■	□	□	■	■	□
Geac SmartStream	■	■	■	■	■	■
GeacVisionShift	□	■	■	■	□	■
Great Plains Software	■	■	■	■	■	■
Hyperion Software Corp.	□	□	□	■	■	■
IET, Inc.	■	■	■	■	■	■
JBA International	■	□	□	■	■	■
JD Edwards	■	■	■	■	■	■
Lawson Software	□	■	■	■	■	■
Macola Inc.	■	■	■	■	■	□
Maconomy NE Inc.	■	□	□	■	■	■
MTX International Inc.	■	■	■	□	□	□
Navison Software US Inc.	□	■	■	■	■	■
OpenPlus International Inc.	□	■	■	■	■	■
Open Systems Inc.	□	□	■	■	■	■
Orange Systems	■	□	■	□	□	■
PeopleSoft	■	■	■	■	■	■
Platinum Software Corporation	■	■	■	■	■	■
PowerCerv	■	□	□	■	□	■
QSP Inc.	□	□	□	■	■	□
Ramco Systems Corp.	■	■	□	■	■	■
Ross Systems Inc.	■	■	■	■	■	■
SAP America Inc.	■	■	■	■	■	■
SAS Institute Inc.	□	□	□	■	□	□
SBT Accounting Systems	■	□	□	■	■	■
Scala North America Inc.	■	□	□	■	■	■
Skylight Systems	□	□	□	■	□	■
Software 2000 Inc.	■	■	■	■	■	■
Solomon Software	■	■	■	■	■	■
SPFC	■	□	■	■	□	□
State of the Art Inc.	□	■	■	■	□	■
SQL Financials	□	■	■	■	□	■
Synon Corporation	□	□	□	■	■	■
Syspro Group	■	■	■	■	□	■
System Software Associates Inc.	■	□	□	■	■	■
Systems Union Inc.	□	□	□	■	■	■
Timeline Inc.	□	□	□	■	□	■
USL Systems	□	■	■	□	□	■
Walker Interactive Systems	□	□	□	■	□	□

Server Support Cross-Reference

	Unix IBM	Unix HP	Unix Sun	MS NT Server	Novell NetWare	IBM OS/400	IBM Mainframe
AccountMate Software	□	□	□	■	□	■	□
Agresso Corp.	■	■	■	■	■	□	□
American Software	■	■	■	■	■	■	■
Apprise Software Inc.	■	■	■	■	■	■	□
Baan Company	■	■	■	■	■	□	□
BHR Software	■	■	■	■	□	□	□
Big Software	□	□	□	■	□	□	□
Carillon Financials Corporation	□	□	□	■	■	□	□
CODA Inc.	■	■	■	■	□	□	□
Computer Associates Int'l Inc.	■	■	■	■	■	■	■
Computron Software Inc.	■	■	■	■	■	□	□
Concepts Dynamic, Inc.	■	■	■	■	■	□	□
Deltek Systems Inc.	■	■	■	■	■	□	□
Design Data Systems	■	■	■	■	■	■	□
FlexiInternational Software Inc.	■	■	■	■	■	■	■
Fourth Shift Corporation	□	□	□	■	□	□	□
Geac SmartStream	■	■	■	■	■	□	□
GeacVisionShift	□	□	□	■	■	□	□
Great Plains Software	□	□	□	■	■	□	□
Hyperion Software Corp.	■	■	■	■	□	□	□
IET, Inc.	■	■	■	■	□	□	□
JBA International	■	■	□	■	□	■	□
JD Edwards	■	■	□	■	□	■	■
Lawson Software	■	■	■	■	□	■	□
Macola Inc.	□	□	□	■	■	□	□
Maconomy NE Inc.	■	■	■	■	□	□	□
MTX International Inc.	□	□	□	■	■	□	□
Navison Software US Inc.	■	■	□	■	□	□	□
OpenPlus International Inc.	■	■	■	■	■	■	■
Open Systems Inc.	□	□	□	■	■	□	□
Orange Systems	■	■	■	■	■	□	□
PeopleSoft	■	■	■	■	■	■	■
Platinum Software Corporation	□	■	■	■	□	□	□
PowerCerv	■	■	■	■	□	□	□
QSP Inc.	■	■	■	□	□	□	■
Ramco Systems Corp.	□	□	□	■	□	□	□
Ross Systems Inc.	■	■	□	■	□	□	□
SAP America Inc.	■	■	■	■	□	■	□
SAS Institute Inc.	■	■	■	■	■	□	□
SBT Accounting Systems	□	□	□	■	■	□	□
Scala North America Inc.	□	■	■	■	■	□	□
Skylight Systems	□	■	■	■	■	□	□
Software 2000 Inc.	□	□	□	□	□	■	□
Solomon Software	□	□	□	■	■	□	□
SPFC	■	■	■	■	■	□	□
State of the Art Inc.	□	□	□	■	□	□	□
SQL Financials	■	■	■	■	□	□	□
Synon Corporation	□	■	□	■	■	■	□
Syspro Group	■	■	■	■	■	□	□
System Software Associates Inc.	■	■	□	□	□	■	□
Systems Union Inc.	■	■	■	■	■	□	□
Timeline Inc.	□	□	□	■	□	□	□
USL Systems	□	■	■	■	■	□	□
Walker Interactive Systems	□	□	□	■	□	□	■

RDBMS Support Cross-Reference

	Oracle 7	Informix	Sybase 11	Microsoft SQL Server	Centura SQLBase	Btrieve/ Scalable SQL	IBM DB2
AccountMate Software	□	□	□	■	□	□	■
Agresso Corp.	■	■	■	■	□	□	□
American Software	■	□	□	□	□	□	■
Apprise Software Inc.	■	□	□	□	□	□	■
Baan Company	■	■	■	■	□	□	□
BHR Software	□	■	□	■	□	□	□
Big Software	□	□	□	□	□	□	□
Carillon Financials Corporation	■	□	■	■	□	□	□
CODA Inc.	■	■	■	■	□	□	□
Computer Associates Int'l Inc.	■	■	□	□	□	□	■
Computron Software Inc.	□	■	□	■	□	□	□
Concepts Dynamic, Inc.	□	■	□	□	□	□	□
Deltek Systems Inc.	■	□	■	■	■	□	□
Design Data Systems	■	□	□	□	□	□	□
FlexiInternational Software Inc.	■	□	■	■	□	□	■
Fourth Shift Corporation	□	□	□	■	□	□	□
Geac SmartStream	□	□	■	■	□	□	□
GeacVisionShift	□	□	□	□	□	□	□
Great Plains Software	□	□	□	■	□	■	□
Hyperion Software Corp.	■	□	■	□	□	□	■
IET, Inc.	■	■	■	■	□	□	■
JBA International	■	□	□	□	□	□	■
JD Edwards	■	□	□	■	□	□	■
Lawson Software	■	■	■	□	□	□	■
Macola Inc.	□	□	□	□	□	■	□
Maconomy NE Inc.	■	□	□	■	□	□	□
MTX International Inc.	□	□	□	■	□	□	□
Navison Software US Inc.	□	□	□	■	□	□	□
OpenPlus International Inc.	■	■	■	■	□	□	■
Open Systems Inc.	□	□	□	■	□	□	□
Orange Systems	■	□	□	□	□	□	■
PeopleSoft	■	■	■	■	■	□	■
Platinum Software Corporation	□	□	■	■	□	□	□
PowerCerv	■	□	■	■	□	□	□
QSP Inc.	■	□	□	□	□	□	■
Ramco Systems Corp.	□	□	□	■	□	□	■
Ross Systems Inc.	■	□	■	□	□	□	□
SAP America Inc.	■	■	□	■	□	□	■
SAS Institute Inc.	■	■	■	■	□	■	■
SBT Accounting Systems	■	■	■	□	□	■	□
Scala North America Inc.	□	■	□	□	□	■	□
Skylight Systems	■	■	■	■	□	□	□
Software 2000 Inc.	□	□	□	□	□	□	■
Solomon Software	□	□	□	■	□	■	□
SPFC	■	□	■	■	■	□	□
State of the Art Inc.	□	□	□	□	□	□	□
SQL Financials	■	□	■	■	□	□	□
Synon Corporation	■	□	□	■	□	□	□
Syspro Group	□	□	□	□	□	□	□
System Software Associates Inc.	■	■	□	■	□	□	■
Systems Union Inc.	■	□	■	■	□	■	□
Timeline Inc.	□	□	□	■	□	□	□
USL Systems	□	□	■	■	□	□	□
Walker Interactive Systems	□	□	□	□	□	□	■

GUI Support Cross-Reference

	MS Windows (32 Bit)	MS Windows (16 Bit)	Apple MacOS	IBM OS/2 Warp	OSF/Motif	OpenLook	Character
AccountMate Software	■	□	□	□	□	□	□
Agresso Corp.	■	■	□	□	□	□	■
American Software	■	■	□	□	□	□	□
Apprise Software Inc.	■	■	□	□	□	□	□
Baan Company	■	□	■	■	□	□	□
BHR Software	■	□	□	□	□	□	■
Big Software	■	□	■	□	□	□	□
Carillon Financials Corp.	■	■	□	□	□	□	□
CODA Inc.	■	■	□	□	□	□	□
Computer Associates Int'l Inc.	■	■	□	□	□	□	■
Computron Software Inc.	■	■	□	□	□	□	■
Concepts Dynamic, Inc.	■	■	□	□	■	□	■
Deltek Systems Inc.	■	■	□	□	□	□	□
Design Data Systems	■	■	■	□	■	□	□
FlexiInternational Software Inc.	■	■	□	□	□	□	□
Fourth Shift Corporation	■	■	□	□	□	□	■
Geac SmartStream	■	■	□	□	□	□	□
GeacVisionShift	■	■	□	□	□	□	□
Great Plains Software	■	■	□	□	□	□	□
Hyperion Software Corp.	■	■	□	□	□	□	□
IET, Inc.	■	■	□	□	□	□	□
JBA International	■	■	□	□	□	□	□
JD Edwards	■	□	□	□	□	□	■
Lawson Software	■	■	□	□	□	□	■
Macola Inc.	■	■	□	□	□	□	□
Maconomy NE Inc.	■	■	■	■	□	□	□
MTX International Inc.	■	■	□	□	□	□	□
Navison Software US Inc.	■	■	□	■	□	□	□
OpenPlus International Inc.	■	■	■	■	■	■	■
Open Systems Inc.	■	■	□	□	□	□	□
Orange Systems	□	□	□	□	□	□	■
PeopleSoft	■	■	□	□	□	□	□
Platinum Software Corp.	■	■	□	□	□	□	□
PowerCerv	■	■	□	□	□	□	□
QSP Inc.	□	■	□	■	□	□	■
Ramco Systems Corp.	■	■	□	□	□	□	□
Ross Systems Inc.	■	■	□	□	□	□	■
SAP America Inc.	■	■	■	■	■	□	□
SAS Institute Inc.	■	■	□	■	□	□	□
SBT Accounting Systems	■	□	□	■	□	□	□
Scala North America Inc.	■	■	□	□	□	□	■
Skylight Systems	■	■	□	□	■	■	■
Software 2000 Inc.	■	■	□	■	□	□	■
Solomon Software	■	■	□	□	□	□	□
SPFC	■	■	□	□	□	□	□
State of the Art Inc.	■	□	□	□	□	□	□
SQL Financials	■	■	□	□	□	□	□
Synon Corporation	■	■	□	□	□	□	■
Syspro Group	■	■	□	□	□	□	■
System Software Assoc. Inc.	■	■	□	□	□	□	□
Systems Union Inc.	■	■	□	□	□	□	■
Timeline Inc.	■	■	□	□	□	□	□
USL Systems	■	■	□	■	□	□	□
Walker Interactive Systems	■	■	□	□	□	□	■

Pricing Cross-Reference

	<US$25K	<US%50K	<US$100K	>US$100K
AccountMate Software	■	□	□	□
Agresso Corp.	■	□	□	□
American Software	□	■	□	□
Apprise Software Inc.	■	□	□	□
Baan Company	□	□	□	■
BHR Software	■	□	□	□
Big Software	■	□	□	□
Carillon Financials Corporation	■	□	□	□
CODA Inc.	□	□	□	■
Computer Associates International Inc.	Please contact vendor			
Computron Software Inc.	□	■	□	□
Concepts Dynamic, Inc.	■	□	□	□
Deltek Systems Inc.	■	□	□	□
Design Data Systems	□	■	□	□
FlexiInternational Software Inc.	□	■	□	□
Fourth Shift Corporation	■	□	□	□
Geac SmartStream	□	□	■	□
GeacVisionShift	■	□	□	□
Great Plains Software	■	□	□	□
Hyperion Software Corp.	□	□	■	■
IET, Inc.	Please contact vendor			
JBA International	■	□	□	□
JD Edwards	□	□	□	■
Lawson Software	■	■	□	□
Macola Inc.	■	□	□	□
Maconomy NE Inc.	□	■	□	□
MTX International Inc.	■	□	□	□
Navison Software US Inc.	■	□	□	□
OpenPlus International Inc.	□	□	■	□
Open Systems Inc.	■	□	□	□
Orange Systems	■	□	□	□
PeopleSoft	□	□	■	□
Platinum Software Corporation	■	□	□	□
PowerCerv	■	□	□	□
QSP Inc.	□	□	■	□
Ramco Systems Corp.	□	■	□	□
Ross Systems Inc.	■	■	■	■
SAP America Inc.	Please contact vendor			
SAS Institute Inc.	□	□	■	□
SBT Accounting Systems	■	□	□	□
Scala North America Inc.	■	□	□	□
Skylight Systems	■	□	□	□
Software 2000 Inc.	□	■	■	■
Solomon Software	■	□	□	□
SPFC	■	□	□	□
State of the Art Inc.	■	□	□	□
SQL Financials	□	□	■	□
Synon Corporation	■	□	□	□
Syspro Group	■	□	□	□
System Software Associates Inc.	■	■	■	■
Systems Union Inc.	■	□	□	□
Timeline Inc.	Please contact vendor			
USL Systems	■	□	□	□
Walker Interactive Systems	□	□	■	■

Sales Channels Cross-Reference

	Direct	VAR Only	Direct and VAR	Big Six/ Outsource Firm
AccountMate Software	☐	■	☐	☐
Agresso Corp.	☐	☐	■	☐
American Software	■	☐	☐	■
Apprise Software Inc.	☐	☐	☐	☐
Baan Company	☐	☐	■	■
BHR Software	☐	☐	■	☐
Big Software	☐	☐	■	☐
Carillon Financials Corporation	☐	☐	■	☐
CODA Inc.	☐	☐	■	☐
Computer Associates International Inc.	☐	☐	■	☐
Computron Software Inc.	☐	☐	■	☐
Concepts Dynamic, Inc.	■	☐	☐	☐
Deltek Systems Inc.	■	☐	☐	☐
Design Data Systems	■	☐	■	■
FlexiInternational Software Inc.	■	☐	☐	■
Fourth Shift Corporation	☐	☐	■	☐
Geac SmartStream	☐	☐	■	☐
GeacVisionShift	☐	☐	■	☐
Great Plains Software	☐	■	☐	■
Hyperion Software Corp.	■	☐	☐	☐
IET, Inc.	■	☐	☐	☐
JBA International	☐	☐	■	☐
JD Edwards	■	☐	■	☐
Lawson Software	☐	☐	■	☐
Macola Inc.	☐	■	☐	☐
Maconomy NE Inc.	☐	☐	■	☐
MTX International Inc.	☐	☐	■	☐
Navison Software US Inc.	☐	■	☐	☐
OpenPlus International Inc.	☐	☐	■	■
Open Systems Inc.	☐	■	☐	☐
Orange Systems	☐	☐	■	☐
PeopleSoft	■	☐	☐	☐
Platinum Software Corporation	☐	☐	■	☐
PowerCerv	☐	☐	■	☐
QSP Inc.	■	☐	☐	☐
Ramco Systems Corp.	■	☐	☐	☐
Ross Systems Inc.	☐	☐	■	☐
SAP America Inc.	☐	☐	■	☐
SAS Institute Inc.	■	☐	☐	☐
SBT Accounting Systems	☐	■	☐	☐
Scala North America Inc.	☐	☐	■	☐
Skylight Systems	☐	☐	■	☐
Software 2000 Inc.	■	☐	☐	■
Solomon Software	☐	■	☐	☐
SPFC	☐	☐	■	☐
State of the Art Inc.	☐	■	☐	☐
SQL Financials	■	☐	☐	☐
Synon Corporation	☐	☐	■	☐
Syspro Group	☐	■	☐	☐
System Software Associates Inc.	■	☐	☐	☐
Systems Union Inc.	☐	☐	■	☐
Timeline Inc.	☐	☐	■	■
USL Systems	☐	☐	■	☐
Walker Interactive Systems	■	☐	☐	☐

Sales and Support Cross-Reference

	USA/ Canada	UK/ Europe	Pacific Rim	Central/ South America
AccountMate Software	■	□	■	■
Agresso Corp.	■	■	□	□
American Software	■	■	■	■
Apprise Software Inc.	■	■	□	□
Baan Company	■	■	■	■
BHR Software	■	□	□	□
Big Software	■	■	■	□
Carillon Financials Corporation	■	□	□	□
CODA Inc.	■	■	■	■
Computer Associates International Inc.	■	■	■	■
Computron Software Inc.	■	■	■	□
Concepts Dynamic, Inc.	■	□	□	□
Deltek Systems Inc.	■	□	□	□
Design Data Systems	■	□	■	■
FlexiInternational Software Inc.	■	■	■	□
Fourth Shift Corporation	■	■	■	■
Geac SmartStream	■	■	■	□
GeacVisionShift	■	□	■	□
Great Plains Software	□	□	□	□
Hyperion Software Corp.	■	■	■	■
IET, Inc.	■	□	□	□
JBA International	■	■	■	■
JD Edwards	■	■	■	□
Lawson Software	■	■	□	□
Macola Inc.	■	■	□	□
Maconomy NE Inc.	■	■	□	□
MTX International Inc.	■	□	□	□
Navison Software US Inc.	■	■	□	□
OpenPlus International Inc.	■	■	■	□
Open Systems Inc.	■	■	■	□
Orange Systems	■	□	□	□
PeopleSoft	■	■	■	■
Platinum Software Corporation	■	■	■	■
PowerCerv	■	■	□	□
QSP Inc.	■	■	□	□
Ramco Systems Corp.	■	■	■	□
Ross Systems Inc.	■	■	■	□
SAP America Inc.	■	■	■	■
SAS Institute Inc.	■	■	■	■
SBT Accounting Systems	■	■	■	■
Scala North America Inc.	■	■	■	■
Skylight Systems	■	■	□	□
Software 2000 Inc.	■	■	■	■
Solomon Software	■	■	■	■
SPFC	■	□	□	□
State of the Art Inc.	■	□	□	□
SQL Financials	■	□	□	□
Synon Corporation	■	■	■	■
Syspro Group	■	■	■	□
System Software Associates Inc.	■	■	■	■
Systems Union Inc.	■	■	■	□
Timeline Inc.	■	■	□	□
USL Systems	■	□	□	□
Walker Interactive Systems	■	■	■	□

Appendix

Reading List

Bochenski, Barbara. Implementing Production Quality Client/Server Systems. New York: John Wiley & Sons, Inc. 1994.

Champny, James and Nohria, Nitin. Fast Forward: The Best Ideas on Managing Business Change. Boston: Harvard Business School Press. 1996.

Cronin, Mary J. The Internet Strategy Handbook. Boston: Harvard Business School Press. 1996.

Fingar, Peter. The Blueprint for Business Objects. New York: SIGS Books. 1996.

Hammer, Michael and Champy, James. Reengineering the Corporation. New York: Harper Collins. 1993.

Harmon, Paul and Hall, Curtis . Intelligent Software Systems Development. New York: John Wiley & Sons, Inc. 1993.

Jacobson, Ivar; Ericsson, Maria; and Jacobson, Agneta. The Object Advantage. New York: ACM Press. 1995.

Kalakota, Ravi and Whinston, Andrew B. Frontiers of Electronic Commerce. New York: Addison-Wesley. 1996.

Keeling, Denis. Corporate Accounting Packages. London: Ovum Ltd. 1995.

Keen, Peter G.W. and Knapp, Ellen M. Every Manager's Guide to Business Processes. Boston: Harvard Business School Press. 1996.

Koulopoulos, Thomas M. The Workflow Imperative. Boston: Delphi Publishing. 1994.

Loomis, Mary E.S. Object Databases: The Essentials. Reading, MA: Addison-Wesley. 1995.

Partridge, Chris. Business Objects: Reengineering for Re-Use. Oxford: Butterworth-Heinemann. 1996.

Pendse, Nigel and Creeth, Richard. The OLAP Report. Norwalk, CT: Business Intelligence. 1995.

Purdum, Jack. Accounting & Finance Developer's Guide. Indianapolis: SAMS Publishing. 1995.

Savage, Charles M. 5th Generation Management. Newton MA: Butterworth-Heinemann. 1996.

Siebel, Thomas M. and Malone, Michael S. Virtual Selling. New York: Simon & Schuster [The Free Press]. 1996.

Taylor, David A. Business Engineering with Object Technology. New York: John Wiley & Sons, Inc. 1995.

Taylor, David A. Object Oriented Information Systems: Planning and Implementation. New York: John Wiley & Sons, Inc. 1992.

Taylor, David A. Object Oriented Technology: A Manager's Guide. Reading, MA: Addison-Wesley. 1990.

Watterson, Karen. Client/Server Technology for Managers. Reading, MA: Addison-Wesley. 1995.

Glossary

Term	Description
4GL	Fourth-generation language — a high-level programming language for fast development and maintenance of applications.
action	The response of a user or a software agent that moves a workflow item forward in a process.
ActiveX	Microsoft technology for delivering and using software applets and components across the Internet and other distributed systems.
agent	A piece of software code programmed to react to certain events without human intervention.
ANSI	American National Standards Institute — a U. S. organization of industry groups that develops standards to facilitate trade and communication. ANSI is a member of the International Standards Organization (ISO).
API	Application programming interface — an interface at the programming level that allows two pieces of software to interact.

applet	A self-contained application that represents just a part of an accounting module, such as a purchase requisition entry function. An applet is small enough to be downloaded across the Internet and run from within desktop Internet browser software.
attachment	A file, such as a text document or a worksheet, that is attached to an e-mail message and that the recipient can open and use.
BOMSIG	The Business Object Management Special Interest Group.
BPR	Business process reengineering — the analysis of business processes in order to determine more effective ways of managing those processes, often through the use of technology.
browser	Software used to view and navigate information accessed via the World Wide Web.
C, C++, Cobol	Programming languages used to build accounting and other applications.
cell	A specific intersection of a row and column in a worksheet.
client	A computer, usually a desktop workstation with local CPU, disk, and RAM, that requests and processes data retrieved from a server located on a network. A client may act as a server to other clients.
column	A discrete data element in a database table.
COM	Microsoft's Component Object Model — a specification for building interoperable software components for use primarily on Microsoft platforms.
conclusion	An event that signifies the conclusion of a workflow, process, or task.
CORBA	The Object Management Group's Common Object Request Broker Architecture, an independent standard for software object management.
Cue Cards	Microsoft Windows pop-up help screens that guide users through a process.

data warehouse	A database used to store a variety of data sourced from many operational systems and optimized for decision-support queries and reporting.
DataLens	Lotus 1-2-3 database access technology for accessing data stored in external databases or file systems.
datamart	A database used to store a focused subset of data sourced from a data warehouse or from one or a handful of closely linked operational systems and optimized for decision-support queries and reporting.
DCOM	Microsoft's Distributed Component Object Model — a specification for building distributed, interoperable software components for use primarily on Microsoft platforms. Regarded in the marketplace as a competitor to CORBA.
DDE	Dynamic Data Exchange — Microsoft technology for transferring data from between desktop applications, such as between spreadsheets and accounting systems.
EDI	Electronic Data Interchange — a specification of common transaction formats for transferring data electronically between heterogenous applications.
EFT	Electronic funds transfer — the mechanism and file formats for transferring payment data electronically between heterogenous applications.
electronic commerce	Managing business transactions electronically; lately used to mean conducting business transactions over the Internet. EDI and EFT are both examples of electronic commerce.
event	A piece of code the system executes in response to a specific user action, such as clicking the mouse, displaying or closing a form, entering or changing data, or pressing the Tab key to move between fields on a form.
extranet	An intranet managed by one party for use for electronic commerce and communication by a specific group of other parties.

FTP	File Transfer Protocol — a standard text-based method of transferring data files across the Internet.
granularity	The level of functional design of a software application.
GUI	Graphical user interface — a visual means of controlling a computer using pictorial objects manipulated by a hand-operated device (e.g., a mouse).
HTML	Hypertext Markup Language — the scripting language used to construct and format the navigable, linked pages of the World Wide Web.
IDE	Integrated development environment — a customization environment built into applications, including some accounting applications.
index	A collection of pointers to specific rows in a database table.
initiation	An event that triggers the start of a workflow, process, or task.
Internet	A wide-area, open-communications infrastructure that hosts data and applications accessible from cross-platform desktop software.
intranet	A private network connected to the Internet but protected from unauthorized access by a security-software-based firewall or proxy server.
ISAM	Indexed sequential access method — a means of storing data sequentially and retrieving it through use of a single index (as opposed to relational databases, which store data nonsequentially and may use many indexes to retrieve it).
item	A workflow item, such as a transaction, form, document, or message, that is subject to routes, rules, and roles.
Java	Sun MicroSystems technology for delivering and using software applets and components across the Internet.

join	A means of relating the data in two or more relational database tables using a key value.
key	One or more columns that uniquely define a row in a relational database table.
LAN	Local-area network — a collection of computers and other peripherals physically connected by wires to enable sharing of data and devices.
log	A record of activity maintained automatically by the database manager and used to help recover the database after a crash.
MAPI	Mail Applications Programming Interface — a Microsoft e-mail protocol.
MDBMS	Multidimensional database management system — a database (often memory-resident) that can store a value for every intersection of its user-defined information dimensions. MDBMSs enable fast data retrieval for queries based on user-specified combinations of dimensions.
MDI	Multiple Document Interface — a Microsoft Windows convention for allowing more than one window to be displayed on screen at a time.
metadata	Data about data, usually a separate set of tables in a database that provide more information about the data stored in a particular application to make it easy for third-party applications to access and manipulate that data.
MHS	Message Handling Service — a Novell e-mail protocol.
middleware	Software used to extract data from source transaction systems and transform it ready for loading into a data warehouse.
MIME	Multipurpose Internet Mail Extensions — a format for encoding data files attached to e-mail messages.
notification	A message sent to inform, alert, or trigger a response from a system user or a system agent.

n-tier	A client/server application that can be deployed across a middle tier of application servers in addition to clients and a database server. The additional tiers allow highly flexible deployment and process- or task-specific scalability.
OAG	Open Applications Group — a group of vendors working to define common interfaces for use in passing data between applications.
ODBC	Open Database Connectivity — a Microsoft middleware product for accessing data stored in relational and nonrelational databases.
ODBMS	Object database management system — a database designed for storing and manipulating software objects that comprise both data (attributes) and the functions (methods) used to manipulate that data.
OLAP	Online Analytical Processing — the processing of complex, user-driven, ad-hoc queries with no predictable pattern.
OLE	Object Linking and Embedding — Microsoft technology for embedding documents, such as spreadsheets, within other documents to create information-rich compound documents.
OLTP	Online transaction processing — the processing of a high volume of short-duration, repetitive transactions such as those initiated at a bank teller machine.
OMG	Object Management Group — an independent group attempting to set standards for the management of software objects.
parallel processing	The ability of a computer to execute operations simultaneously ("in parallel") rather than sequentially.
partitioning	A way of designing a software application so its functionality can be distributed over separate tiers (platforms), making it easier to deploy the application over a variety of client- or server-centric configurations.

pivot table	A method for rotating row and column dimensions in Microsoft Excel to see new views of the data in the worksheet.
process	A series of logically linked tasks.
proxy server	A server that provides security and other services to protect intranet database and application servers from access by unauthorized users.
publish and subscribe	A form of data replication in which a database is nominated as the "publisher" of data and one or more databases are nominated as "subscribers" to data from that database.
query	An SQL statement for retrieving data from a relational database.
RDBMS	Relational database management system — a database that stores and manages data using the methodology originally proposed by Dr. E.F. Codd.
replication	The automated copying of data from a source database to one or more target databases.
role	A skillset or organizational position required to perform a specific task.
route	A transfer of information from a sending role to a receiving role dependent on a specific set of rules.
row	A unique collection of columns in a table.
RPC	Remote procedure call — a usually simple request from a client application to run a process on a remote server.
rule	A conditional parameter that determines an action to be taken as part of a workflow transaction.
scalability	The ability to adapt a system to accommodate more transactions, more concurrent users, or additional functionality.
server	A computer dedicated to a specific task (e.g., running a database management system) that provides services to multiple clients connected via a network. A server may act as a client to other servers.

SMTP	Simple Mail Transfer Protocol — an Internet e-mail protocol.
SOHO	Single office/home office — usually a small business operated from the owner's home.
source code	The actual program code of which an application is composed.
SQL	Structured Query Language — a computer language used to create, manipulate, and access relational databases.
state	The current status of a workflow item relative to the process of which it is a part.
store and forward	A mechanism whereby message data is stored locally until a connection is made with the remote computer that allows the message to be forwarded to the designated recipient.
stored procedure	A piece of code that is stored in a database and that applications can execute via a remote procedure call to carry out a task on a server.
table	A collection of data rows with one or more indexes in a relational database.
task	A finite set of actions with a defined initiation and conclusion that can usually be timed. Also known as an activity.
thick client	A client with the high level of resources (CPU, RAM, disk) needed to support client-centric processing.
thin client	A client that has a low level of resources (CPU, RAM, disk) because the applications it uses are server-centric.
three-tier	Describes a client/server application that can use an application server between the client and the database server to provide more flexible deployment and scalability than a two-tier application.
tier	The layers of application functionality in a partitioned client/server application that can usually be

	deployed on separate physical computers for better scalability.
trigger	A piece of code in a database that automatically reacts to certain events without user intervention.
TWAIN	A commercial standard for digitizing information via a scanning device.
two-tier	Describes a conventional client/server application that runs partly on a client and partly on a server.
URL	Universal Resource Locator — the address of a site on the World Wide Web.
VBA	Visual Basic for Applications — a Microsoft derivative of the popular Visual Basic programming lanaguage that allows applications running on Microsoft platforms to interoperate.
view	A "virtual" table that contains selected data from one or more tables in a relational database.
VIM	Vendor Independent Messaging — a Lotus Development Corporation e-mail protocol.
Visual Basic	A popular Microsoft programming language used by accounting software vendors to build graphical client application interfaces.
WAN	Wide-area network — a collection of computers and other peripherals connected by telecommunication lines to enable sharing of data and devices.
Web browser	A program that provides a means of accessing text, graphics, and applications stored on World Wide Web servers.
WfMC	Workflow Management Coalition — a nonprofit, Brussels-based standards organization focused on defining workflow standards.
widget	A specific graphical control, such as a button or a box, used by a GUI to accept or respond to actions by the user.

wizard | A Microsoft Windows pop-up help screen that prompts the user for input and guides the user down a particular process path depending on that input.

workbook | A tabbed "container" for storing multiple worksheets in Microsoft Excel. Clicking the tab with the mouse displays the worksheet linked to the tab.

workflow | The definable route of an item through one or more processes from start to finish where each event that affects the item and the consequences of those events are fully defined.

WWW | The World Wide Web — a service on the Internet.

WYSIWYG | "What you see is what you get," meaning that what the user sees on screen is what the data looks like when printed.

Index

New Books in the Duke Press Library

MIGRATING TO WINDOWS NT 4.0

By Sean Daily

A comprehensive yet concise guide to the significant changes users will encounter as they make the move to Windows NT 4.0. Includes a wealth of tips and techniques. 9 chapters, 450 pages.

POWERING YOUR WEB SITE WITH WINDOWS NT SERVER

By Nik Simpson

Explores the tools necessary to establish a presence on the Internet or on an internal corporate intranet using WWW technology and Windows NT Server. 500 pages. CD included.

MICROSOFT EXCHANGE — UP AND RUNNING!

By Bill Kilcullen

A practical guide to incorporating the Exchange model. This book is the link between technical manuals and everyday concerns faced by professionals charged with the task of implementing and managing a complex messaging system. 300 pages. CD included.

THE MICROSOFT EXCHANGE USER'S HANDBOOK

By Sue Mosher

A must-have, complete guide for users who need to know how to set up and use all the features of the Microsoft Exchange client product. 730 pages. CD included.

THE ADMINISTRATOR'S GUIDE TO MICROSOFT SQL SERVER 6.5

By Kevin Cox and William Jones

Delivers expert technical advice, practical management guidelines, and an in-depth look at the database administration aspects of the Microsoft SQL Server 6.5 product. 450 pages.

THE MICROSOFT EXCHANGE SERVER INTERNET MAIL CONNECTOR

By Spyros Sakellariadis

Presents everything you need to know about how to plan, install, and configure the servers in your Exchange environment to achieve the Internet connectivity users demand. 200 pages.

DEVELOPING YOUR AS/400 INTERNET STRATEGY

By Alan Arnold

Addresses the issues unique to deploying your AS/400 on the Internet. Includes procedures for configuring AS/400 TCP/IP and information about which client and server technologies the AS/400 supports natively. Don't put precious corporate data and systems in harm's way. Arnold shows you how to reconcile the AS/400 security-conscious mindset with the less secure philosophy of the Internet community. This enterprise-class tutorial evaluates the AS/400 as an Internet server and teaches you how to design, program, and manage your Web home page. 225 pages.

Also Published by Duke Press

THE A TO Z OF EDI

By Nahid M. Jilovec

Electronic Data Interchange (EDI) can help reduce administrative costs, accelerate information processing, ensure data accuracy, and streamline business procedures. Here's a comprehensive guide to EDI to help in planning, startup, and implementation. The author reveals all the benefits, challenges, standards, and implementation secrets gained through extensive experience. She shows how to evaluate your business procedures, select special hardware and software, establish communications requirements and standards, address audit issues, and employ the legal support necessary for EDI activities. 263 pages.

APPLICATION DEVELOPER'S HANDBOOK FOR THE AS/400

Edited by Mike Otey, a **NEWS/400** *technical editor*

Explains how to effectively use the AS/400 to build reliable, flexible, and efficient business applications. Contains RPG/400 and CL coding examples and tips, and provides both step-by-step instructions and handy reference material. Includes diskette. 768 pages, 48 chapters.

AS/400 DISK SAVING TIPS & TECHNIQUES

By James R. Plunkett

Want specific help for cleaning up and maintaining your disks? Here are more than 50 tips, plus design techniques for minimizing your disk usage. Each tip is completely explained with the "symptom," the problem, and the technique or code you need to correct it. 72 pages.

AS/400 SUBFILES IN RPG

On the AS/400, subfiles are powerful and easy to use, and with this book you can start working with subfiles in just a few hours — no need to wade through page after page of technical jargon. You'll start with the concept behind subfiles, then discover how easy they are to program. The book contains all of the DDS subfile keywords announced in V2R3 of OS/400. Five complete RPG subfile programs are included, and the book comes complete with a 3.5" PC diskette containing all those programs plus DDS. The book is an updated version of the popular *Programming Subfiles in RPG/400*. 200 pages, 4 chapters.

C FOR RPG PROGRAMMERS

By Jennifer Hamilton, a **NEWS/400** *author*

Written from the perspective of an RPG programmer, this book includes side-by-side coding examples written in both C and RPG clear identification of unique C constructs, and a comparison of RPG op-codes to equivalent C concepts. Includes many tips and examples covering the use of C/400. 292 pages, 23 chapters.

CL BY EXAMPLE

By Virgil Green

CL by Example gives programmers and operators more than 850 pages of practical information you can use in your day-to-day job. It's full of application examples, tips, and techniques, along with a sprinkling of humor. The examples will speed you through the learning curve to help you become a more proficient, more productive CL programmer. 864 pages, 12 chapters.

CLIENT ACCESS TOKEN-RING CONNECTIVITY

By Chris Patterson

Attaching PCs to AS/400s via a Token-Ring can become a complicated subject — when things go wrong, an understanding of PCs, the Token-Ring, and OS/400 is often required. *Client Access Token-Ring Connectivity* details all that is required in these areas to successfully maintain and troubleshoot a Token-Ring network. The first half of the book introduces the Token-Ring and describes the Client Access communications architecture, the Token-Ring connection from both the PC side and the AS/400 side, and the Client Access applications. The second half provides a useful guide to Token-Ring management, strategies for Token-Ring error identification and recovery, and tactics for resolving Client Access error messages. 125 pages, 10 chapters.

COMMON-SENSE C

Advice and warnings for C and C++ programmers

By Paul Conte, a **NEWS/400** *technical editor*

C programming language has its risks; this book shows how C programmers get themselves into trouble, includes tips to help you avoid C's pitfalls, and suggests how to manage C and C++ application development. 100 pages, 9 chapters.

CONTROL LANGUAGE PROGRAMMING FOR THE AS/400

By Bryan Meyers and Dan Riehl, **NEWS/400** *technical editors*

This comprehensive CL programming textbook offers students up-to-the-minute knowledge of the skills they will need in today's MIS environment. Progresses methodically from CL basics to more complex processes and concepts, guiding readers toward a professional grasp of CL programming techniques and style. 512 pages, 25 chapters.

DDS BY EXAMPLE

By R S Tipton

DDS by Example provides detailed coverage of the creation of physical files, field reference files, logical files, display files, and printer files. It includes more than 300 real-life examples, including examples of physical files, simple logical files, multi-format logical files, dynamic selection options, coding subfiles, handling overrides, creating online help, creating reports, and coding windows. 360 pages, 4 chapters.

DDS PROGRAMMING FOR DISPLAY & PRINTER FILES

By James Coolbaugh

Offers a thorough, straightforward explanation of how to use Data Description Specifications (DDS) to program display files and printer files. Covers basic to complex tasks using DDS functions. The author uses DDS programming examples for CL and RPG extensively throughout the book, and you can put these examples to use immediately. Focuses on topics such as general screen presentations, the A specification, defining data on the screen, record-format and field definitions, defining data fields, using indicators, data and text attributes, cursor and keyboard control, editing data, validity checking, response keywords, and function keys. A complimentary diskette includes all the source code presented in the book. 446 pages, 13 chapters.

DATABASE DESIGN AND PROGRAMMING FOR DB2/400

By Paul Conte

This textbook is the comprehensive guide for creating flexible and efficient application databases in DB2/400. The author shows you everything you need to know about physical and logical file DDS, SQL/400, and RPG IV and COBOL/400 database programming. Clear explanations illustrated by a wealth of examples, including complete RPG IV and COBOL/400 programs, demonstrate efficient database programming and error handling with both DDS and SQL/400. Each programming chapter includes a specific list of "Coding Suggestions" that will help you write faster and more maintainable code. In addition, the author provides an extensive section on practical database design for DB2/400. This is the most complete guide to DB2/400 design and programming available anywhere. Approx. 772 pages, 19 chapters.

DESKTOP GUIDE TO THE S/36

By Mel Beckman, Gary Kratzer, and Roger Pence, **NEWS/400** *technical editors*
This definitive S/36 survival manual includes practical techniques to supercharge your S/36, including ready-to-use information for maximum system performance tuning, effective application development, and smart Disk Data Management. Includes a review of two popular Unix-based S/36 work-alike migration alternatives. Diskette contains ready-to-run utilities to help you save machine time and implement power programming techniques such as External Program Calls. 387 pages, 21 chapters.

THE ESSENTIAL GUIDE TO CLIENT ACCESS FOR DOS EXTENDED

By John Enck, Robert E. Anderson, and Michael Otey
The Essential Guide to Client Access for DOS Extended contains key insights and need-to-know technical information about Client Access for DOS Extended, IBM's strategic AS/400 product for DOS and Windows client/server connectivity. This book provides background information about the history and architecture of Client Access for DOS Extended; fundamental information about how to install and configure Client Access; and advanced information about integrating Client Access with other types of networks, managing how Client Access for DOS Extended operates under Windows, and developing client/server applications with Client Access. Written by industry experts based on their personal and professional experiences with Client Access, this book can help you avoid time-consuming pitfalls that litter the path of AS/400 client/server computing. 430 pages, 12 chapters.

ILE: A FIRST LOOK

By George Farr and Shailan Topiwala
This book begins by showing the differences between ILE and its predecessors, then goes on to explain the essentials of an ILE program — using concepts such as modules, binding, service programs, and binding directories. You'll discover how ILE program activation works and how ILE works with its predecessor environments. The book covers the new APIs and new debugging facilities and explains the benefits of ILE's new exception-handling model. You also get answers to the most commonly asked questions about ILE. 183 pages, 9 chapters.

IMPLEMENTING AS/400 SECURITY, SECOND EDITION

A practical guide to implementing, evaluating, and auditing your AS/400 security strategy
By Wayne Madden, a **NEWS/400** *technical editor*
Concise and practical, this second edition brings together in one place the fundamental AS/400 security tools and experience-based recommendations that you need and also includes specifics on the latest security enhancements available in OS/400 Version 3 Release 1. Completely updated from the first edition, this is the only source for the latest information about how to protect your system against attack from its increasing exposure to hackers. 389 pages, 16 chapters.

INSIDE THE AS/400
An in-depth look at the AS/400's design, architecture, and history

By Frank G. Soltis

The inside story every AS/400 developer has been waiting for, told by Dr. Frank G. Soltis, IBM's AS/400 chief architect. Never before has IBM provided an in-depth look at the AS/400's design, architecture, and history. This authoritative book does just that — and also looks at some of the people behind the scenes who created this revolutionary system for you. Whether you are an executive looking for a high-level overview or a "bit-twiddling techie" who wants all the details, *Inside the AS/400* demystifies this system, shedding light on how it came to be, how it can do the things it does, and what its future may hold — especially in light of its new PowerPC RISC processors. 475 pages, 12 chapters.

INTRODUCTION TO AS/400 SYSTEM OPERATIONS

by Patrice Gapen and Heidi Rothenbuehler

Here's the textbook that covers what you need to know to become a successful AS/400 system operator. System operators typically help users resolve problems, manage printed reports, and perform regularly scheduled procedures. *Introduction to AS/400 System Operations* introduces a broad range of topics, including system architecture; DB2/400 and Query; user interface and Operational Assistant; managing jobs and printed reports; backup and restore; system configuration and networks; performance; security; and Client Access (PC Support).

   The information presented here covers typical daily, weekly, and monthly AS/400 operations using V3R1M0 of the OS/400 operating system. You can benefit from this book even if you have only a very basic knowledge of the AS/400. If you know how to sign on to the AS/400, and how to use the function keys, you're ready for the material in this book. 234 pages, 10 chapters.

AN INTRODUCTION TO COMMUNICATIONS FOR THE AS/400, SECOND EDITION

By John Enck and Ruggero Adinolfi

This second edition has been revised to address the sweeping communications changes introduced with V3R1 of OS/400. As a result, this book now covers the broad range of AS/400 communications technology topics, ranging from Ethernet to X.25, and from APPN to AnyNet. The book presents an introduction to data communications and then covers communications fundamentals, types of networks, OSI, SNA, APPN, networking roles, the AS/400 as host and server, TCP/IP, and the AS/400-DEC connection. 210 pages, 13 chapters.

JIM SLOAN'S CL TIPS & TECHNIQUES

By Jim Sloan, developer of QUSRTOOL's TAA Tools

Written for those who understand CL, this book draws from Jim Sloan's knowledge and experience as a developer for the S/38 and the AS/400, and his creation of QUSRTOOL's TAA tools, to give you tips that can help you write better CL programs and become more productive. Includes more than 200 field-tested techniques, plus exercises to help you understand and apply many of the techniques presented. 564 pages, 30 chapters.

MASTERING AS/400 PERFORMANCE

by Alan Arnold, Charly Jones, Jim Stewart, and Rick Turner

If you want more from your AS/400 — faster interactive response time, more batch jobs completed on time, and maximum use of your expensive resources — this book is for you. In *Mastering AS/400 Performance*, the experts tell you how to measure, evaluate, and tune your AS/400's performance. From the authors' experience in the field, they give you techniques for improving performance beyond simply buying additional hardware. Learn the techniques, gain the insight, and help your company profit from the experience of the top AS/400 performance professionals in the country. 259 pages, 14 chapters.

MASTERING THE AS/400
A practical, hands-on guide
By Jerry Fottral

This introductory textbook to AS/400 concepts and facilities has a utilitarian approach that stresses student participation. A natural prerequisite to programming and database management courses, it emphasizes mastery of system/user interface, member-object-library relationship, utilization of CL commands, and basic database and program development utilities. Also includes labs focusing on essential topics such as printer spooling; library lists; creating and maintaining physical files; using logical files; using CL and DDS; working in the PDM environment; and using SEU, DFU, Query, and SDA. 484 pages, 12 chapters.

OBJECT-ORIENTED PROGRAMMING FOR AS/400 PROGRAMMERS
By Jennifer Hamilton, a **NEWS/400** *author*

Explains basic OOP concepts such as classes and inheritance in simple, easy-to-understand terminology. The OS/400 object-oriented architecture serves as the basis for the discussion throughout, and concepts presented are reinforced through an introduction to the C++ object-oriented programming language, using examples based on the OS/400 object model. 114 pages, 14 chapters.

PERFORMANCE PROGRAMMING — MAKING RPG SIZZLE
By Mike Dawson, CDP

Mike Dawson spent more than two years preparing this book — evaluating programming options, comparing techniques, and establishing benchmarks on thousands of programs. "Using the techniques in this book," he says, "I have made program after program run 30%, 40%, even 50% faster." To help you do the same, Mike gives you code and benchmark results for initializing and clearing arrays, performing string manipulation, using validation arrays with look-up techniques, using arrays in arithmetic routines, and a lot more. 257 pages, 8 chapters.

POWER TOOLS FOR THE AS/400, VOLUMES I AND II
Edited by Frederick L. Dick and Dan Riehl

NEWS 3X/400's Power Tools for the AS/400 is a two-volume reference series for people who work with the AS/400. *Volume I* (originally titled *AS/400 Power Tools*) is a collection of the best tools, tips, and techniques published in *NEWS/34-38* (pre-August 1988) and *NEWS 3X/400* (August 1988 through October 1991) that are applicable to the AS/400. *Volume II* extends this original collection by including material that appeared through 1994. Each book includes a diskette that provides load-and-go code for easy-to-use solutions to many everyday problems. *Volume I*: 709 pages, 24 chapters; V*olume II*: 702 pages, 14 chapters.

PROGRAMMING IN RPG IV
By Judy Yaeger, Ph.D., a **NEWS/400** *technical editor*

This textbook provides a strong foundation in the essentials of business programming, featuring the newest version of the RPG language: RPG IV. Focusing on real-world problems and down-to-earth solutions using the latest techniques and features of RPG, this book provides everything you need to know to write a well-designed RPG IV program. Each chapter includes informative, easy-to-read explanations and examples as well as a section of thought-provoking questions, exercises, and programming assignments.

Four appendices and a handy, comprehensive glossary support the topics presented throughout the book. An instructor's kit is available. 450 pages, 13 chapters.

PROGRAMMING IN RPG/400, SECOND EDITION

By Judy Yaeger, Ph.D., a **NEWS/400** *technical editor*

This second edition refines and extends the comprehensive instructional material contained in the original textbook and features a new section that introduces externally described printer files, a new chapter that highlights the fundamentals of RPG IV, and a new appendix that correlates the key concepts from each chapter with their RPG IV counterparts. Includes everything you need to learn how to write a well-designed RPG program, from the most basic to the more complex, and each chapter includes a section of questions, exercises, and programming assignments that reinforce the knowledge you have gained from the chapter and strengthen the groundwork for succeeding chapters. An instructor's kit is available. 464 pages, 14 chapters.

PROGRAMMING SUBFILES IN COBOL/400

By Jerry Goldson

Learn how to program subfiles in COBOL/400 in a matter of hours! This powerful and flexible programming technique no longer needs to elude you. You can begin programming with subfiles the same day you get the book. You don't have to wade through page after page, chapter after chapter of rules and parameters and keywords. Instead, you get solid, helpful information and working examples that you can apply to your application programs right away. 204 pages, 5 chapters.

THE QUINTESSENTIAL GUIDE TO PC SUPPORT

By John Enck, Robert E. Anderson, Michael Otey, and Michael Ryan

This comprehensive book about IBM's AS/400 PC Support connectivity product defines the architecture of PC Support and its role in midrange networks, describes PC Support's installation and configuration procedures, and shows you how you can configure and use PC Support to solve real-life problems. 345 pages, 11 chapters.

RPG ERROR HANDLING TECHNIQUE

Bulletproofing Your Applications

By Russell Popeil

RPG Error Handling Technique teaches you the skills you need to use the powerful tools provided by OS/400 and RPG to handle almost any error from within your programs. The book explains the INFSR, INFDS, PSSR, and SDS in programming terms, with examples that show you how all these tools work together and which tools are most appropriate for which kind of error or exception situation. It continues by presenting a robust suite of error/exception handling techniques within RPG programs. Each technique is explained in an application setting, using both RPG III and RPG IV code. 164 pages, 5 chapters.

RPG IV BY EXAMPLE

By George Farr and Shailan Topiwala

RPG IV by Example addresses the needs and concerns of RPG programmers at any level of experience. The focus is on RPG IV in a practical context that lets AS/400 professionals quickly grasp what's new without dwelling on the old. Beginning with an overview of RPG IV specifications, the authors prepare the way for examining all the features of the new version of the language. The chapters that follow explore RPG IV further with practical, easy-to-use applications. 500 pages, 15 chapters.

RPG IV JUMP START, SECOND EDITION
Moving ahead with the new RPG

By Bryan Meyers, a **NEWS/400** *technical editor*

In this second edition of *RPG IV Jump Start*, Bryan Meyers has added coverage for new releases of the RPG IV compiler (V3R2, V3R6, and V3R7) and amplified the coverage of RPG IV's participation in the integrated language environment (ILE). As in the first edition, he covers RPG IV's changed and new specifications and data types. He presents the new RPG from the perspective of a programmer who already knows the old RPG, pointing out the differences between the two and demonstrating how to take advantage of the new syntax and function. 204 pages, 16 chapters.

RPG/400 INTERACTIVE TEMPLATE TECHNIQUE

By Carson Soule, CDP, CCP, CSP

Here's an updated version of Carson Soule's *Interactive RPG/400 Programming*. The book shows you time-saving, program-sharpening concepts behind the template approach, and includes all the code you need to build one perfect program after another. These templates include code for cursor-sensitive prompting in DDS, for handling messages in resident RPG programs, for using the CLEAR opcode to eliminate hard-coded field initialization, and much more. There's even a new select template with a pop-up window. 258 pages, 10 chapters.

S/36 POWER TOOLS

Edited by Chuck Lundgren, a **NEWS/400** *technical editor*

Winner of an STC Award of Achievement in 1992, this book contains five years' worth of articles, tips, and programs published in *NEWS 3X/400* from 1986 to October 1990, including more than 280 programs and procedures. Extensively cross-referenced for fast and easy problem solving, and complete with diskette containing all the programming code. 738 pages, 20 chapters.

STARTER KIT FOR THE AS/400, SECOND EDITION
An indispensable guide for novice to intermediate AS/400 programmers and system operators

By Wayne Madden, a **NEWS/400** *technical editor*
with contributions by Bryan Meyers, Andrew Smith, and Peter Rowley

This second edition contains updates of the material in the first edition and incorporates new material to enhance its value as a resource to help you learn important basic concepts and nuances of the AS/400 system. New material focuses on installing a new release, working with PTFs, AS/400 message handling, working with and securing printed output, using operational assistant to manage disk space, job scheduling, save and restore basics, and more basic CL programming concepts. Optional diskette available. 429 pages, 33 chapters.

SUBFILE TECHNIQUE FOR RPG/400 PROGRAMMERS

By Jonathan Yergin, CDP, and Wayne Madden

Here's the code you need for a complete library of shell subfile programs: RPG/400 code, DDS, CL, and sample data files. There's even an example for programming windows. You even get some "whiz bang" techniques that add punch to your applications. This book explains the code in simple, straightforward style and tells you when each technique should be used for best results. 326 pages, 11 chapters, 3.5" PC diskette included.

TECHNICAL REFERENCE SERIES

Edited by Bryan Meyers, a **NEWS/400** *technical editor*

Written by experts — such as John Enck, Bryan Meyers, Julian Monypenny, Roger Pence, Dan Riehl — these unique desktop guides put the latest AS/400 applications and techniques at your fingertips. These "just-do-it" books (featuring wire-o binding to open flat at every page) are priced so you can keep your personal set handy. Optional online Windows help diskette available for each book.

Desktop Guide to CL Programming

By Bryan Meyers, a **NEWS/400** *technical editor*

This first book of the **NEWS/400** *Technical Reference Series* is packed with easy-to-find notes, short explanations, practical tips, answers to most of your everyday questions about CL, and CL code segments you can use in your own CL programming. Complete "short reference" lists every command and explains the most-often-used ones, along with names of the files they use and the MONMSG messages to use with them. 205 pages, 36 chapters.

Desktop Guide to AS/400 Programmers' Tools

By Dan Riehl, a **NEWS/400** *technical editor*

This second book of the **NEWS/400** *Technical Reference Series* gives you the "how-to" behind all the tools included in *Application Development ToolSet/400* (ADTS/400), IBM's Licensed Program Product for Version 3 of OS/400; includes Source Entry Utility (SEU), Programming Development Manager (PDM), Screen Design Aid (SDA), Report Layout Utility (RLU), File Compare/Merge Utility (FCMU), and Interactive Source Debugger. Highlights topics and functions specific to Version 3 of OS/400. 266 pages, 30 chapters.

Desktop Guide to DDS

By James Coolbaugh

This third book of the **NEWS/400** *Technical Reference Series* provides a complete reference to all DDS keywords for physical, logical, display, printer, and ICF files. Each keyword is briefly explained, with syntax rules and examples showing how to code the keyword. All basic and pertinent information is provided for quick and easy access. While this guide explains every parameter for a keyword, it doesn't explain every possible exception that might exist. Rather, the guide includes the basics about what each keyword is designed to accomplish. The *Desktop Guide to DDS* is designed to give quick, "at your fingertips" information about every keyword — with this in hand, you won't need to refer to IBM's bulky *DDS Reference* manual. 132 pages, 5 major sections.

Desktop Guide to RPG/400

By Roger Pence and Julian Monypenny, **NEWS/400** *technical editors*

This fourth book in the *Technical Reference Series* provides a variety of RPG templates, subroutines, and copy modules, sprinkled with evangelical advice that will help you write robust and effective RPG/400 programs. Highlights of the information provided include string-handling routines, numeric editing routines, date routines, error-handling modules, tips for using OS/400 APIs with RPG/400, and interactive programming techniques. For all types of RPG projects, this book's tested and ready-to-run building blocks will easily snap into your RPG. The programming solutions provided here would otherwise take you days or even weeks to write and test. 211 pages, 28 chapters.

Desktop Guide to Creating CL Commands

By Lynn Nelson

In this most recent book in the *Technical Reference Series*, author Lynn Nelson shows you how to create your own CL commands with the same functionality and power as the IBM commands you use every day, including automatic parameter editing, all the function keys, F4 prompt for values, expanding lists of values, and conditional prompting. After you have read this book, you can write macros for the operations you do over and over every day or write application commands that prompt users for essential information. Whether you're in operations or programming, don't miss this opportunity to enhance your career-building skills. 164 pages, 14 chapters.

UNDERSTANDING BAR CODES

By James R. Plunkett

One of the most important waves of technology sweeping American industry is the use of bar coding to capture and track data. The wave is powered by two needs: the need to gather information in a more accurate and timely manner and the need to track that information once it is gathered. Bar coding meets these needs and provides creative and cost-effective solutions for many applications. With so many leading-edge technologies, it can be difficult for IS professionals to keep up with the concepts and applications they need to make solid decisions. This book gives you an overview of bar code technology including a discussion of the bar codes themselves, the hardware that supports bar coding, how and when to justify and then implement a bar code application, plus examples of many different applications and how bar coding can be used to solve problems. 70 pages.

USING QUERY/400

By Patrice Gapen and Catherine Stoughton

This textbook, designed for any AS/400 user from student to professional with or without prior programming knowledge, presents Query as an easy and fast tool for creating reports and files from AS/400 databases. Topics are ordered from simple to complex and emphasize hands-on AS/400 use; they include defining database files to Query, selecting and sequencing fields, generating new numeric and character fields, sorting within Query, joining database files, defining custom headings, creating new database files, and more. Instructor's kit available. 92 pages, 10 chapters.

USING VISUAL BASIC WITH CLIENT ACCESS APIs

By Ron Jones

This book is for programmers who want to develop client/server solutions on the AS/400 and the personal computer. Whether you are a VB novice or a VB expert, you will gain by reading this book because it provides a thorough overview of the principles and requirements for programming in Windows using VB. Companion diskettes contain source code for all the programming projects referenced in the book, as well as for numerous other utilities and programs. All the projects are compatible with Windows 95 and VB 4.0. 680 pages, 13 chapters.

Subscribe Now No Risk!

... for timely information that helps you every day in your job...delivered to your desk each month. Don't miss a single issue!

Try it at no risk – Satisfaction guaranteed!

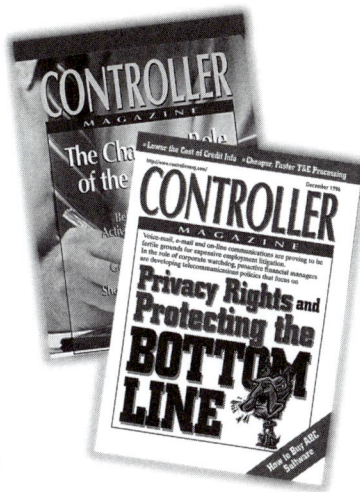

❏ **Yes,** enter my subscription to *Controller Magazine* for 1 year (12 issues) for $59 and bill me. If I change my mind, I'll return your invoice marked "cancel" and keep my first issue **FREE**.

Name _____

Title _____ Phone _____

Company _____

Address _____

City, State, Zip _____

E-mail _____

Price valid in U.S. only.

❏ Please send me information about new Duke Press books on the subject of

_____

_____

Copy this page and then mail to
Duke Press • 221 E. 29th Street • Loveland, CO 80538
or fax it to (970)663-4007
Browse and Shop at www.dukepress.com

[97ACCTSOFT]